A Second Course
in Elementary
Differential Equations

A Second Course in Elementary Differential Equations

Paul Waltman
Emory University, Atlanta

Academic Press, Inc.
(Harcourt Brace Jovanovich, Publishers)
Orlando   San Diego   San Francisco   New York
London   Toronto   Montreal   Sydney   Tokyo   São Paulo

Academic Press, Inc.
Orlando, Florida 32887

United Kingdom Edition Published by Academic Press, Inc.
(London) Ltd., 24/28 Oval Road, London NW1 7DX

ISBN: 0-12-733910-8
Library of Congress Catalog Card Number: 85-70251

Printed in the United States of America

TO RUTH
*For her patience
and understanding*

Contents

3

Existence Theory   161

4

Boundary Value Problems   195

Preface

The once-standard course in elementary differential equations has undergone a considerable transition in the past two decades. In an effort to bring the subject to a wider audience, a gentle introduction to differential equations is frequently incorporated into the basic calculus course – particularly in the last semester of a calculus-for-engineers course – or given separately, but strictly as a problem-solving course. With this approach, students frequently learn how to solve constant coefficient scalar differential equations and little else. If systems of differential equations are treated, the treatment is usually incomplete. Laplace transform techniques, series solutions, or some existence theorems are sometimes included in such a course, but seldom is any of the flavor of modern differential equations imparted to the student. Graduates of such a course are often ill-equipped to take the next step in their education, which all too frequently is a graduate-level differential equations course with considerable analytical prerequisites.

Even when a "good" elementary course in ordinary differential equations is offered, the student who needs to know more sophisticated topics may find his or her way to further study blocked by the need to first study real and complex variables, functional analysis, and so on. Yet many of the more modern topics can be taught along with a marginal amount of the necessary analysis. This book is for a course directed toward students who need to know more about ordinary differential equations; who, perhaps as mathematics or physics students, have not yet had the time to study sufficient analysis to be able to master an honest course, or who, perhaps as biologists, engineers, economists, and so on, cannot take the necessary time to master the prerequisites for a graduate course in mathematics but who need to know more of the subject.

This book, then, is a second course in (elementary) ordinary differential equations, a course that may be taken by those with minimal – but not zero – preparation in ordinary differential equations, and yet which treats some topics from a sufficiently advanced point of view so that even those students with good preparation will find something of interest and value. I have taught the subject at this level at Arizona State University and The University of Iowa to classes with extremely varied backgrounds and levels of mathematical sophistication with very satisfactory results. This book is the result of those courses.

Before describing the contents, I wish to further emphasize a topic alluded to above. For some students from other disciplines, this may be the only analysis course they will see in their mathematical education. Thus, whenever possible, basic real analysis, as well as differential equations, is taught. The concepts of analysis are brought into play wherever possible — ideas such as norms, metric spaces, completeness, inner products, asymptotic behavior, and so on, are introduced in the natural setting of a need to solve, or to set, a problem in differential equations. For example, metric spaces could be avoided in the proof of the existence theorem but they are deliberately used, because the idea of an abstract space is important in much of applied mathematics and it can be introduced easily and naturally in the context of very simple operators.

The book has applications as well. However, rather than tossing in trivial applications of dubious practical use, few, but detailed, applications are treated, with some attention given to the mathematical modeling that leads to the equation. By and large, however, the book is about applicable, rather than truly applied, mathematics.

Chapter 1 gives a thorough treatment of linear systems of differential equations. Necessary concepts from linear algebra are reviewed and the basic theory is presented. The constant coefficient case is presented in detail, and all cases are treated, even that of repeated eigenvalues. The novelty here is the treatment of the case of the nondiagonalizable coefficient matrix without the use of the Jordan form. I have had good results substituting the Putzer algorithm — it gives a computational procedure that students can master. This part of the course, which goes rather quickly, is computational and helps pull students with different backgrounds to the same level. Topics in stability of systems and the case of periodic coefficients are included for a more able class.

Chapter 2 is the heart of the course, where the ideas of stability and qualitative behavior are developed. Two-dimensional linear systems form the starting point — the phase plane concepts. Polar coordinate techniques play a role here. Liapunov stability and elementary ideas from dynamic systems are treated. Limit cycles appear here as an example of a truly nonlinear phenomenon. In a real sense, this is "applied topology" and some topological ideas are gently introduced. The Poincaré-Bendixson theorem is stated and its significance discussed. Of course, proofs at this stage are too difficult to present; so, if the first section can be described as computational, then this section is geometrical and intuitive.

Chapter 3 presents existence and uniqueness theorems in a rigorous way. Not all students will profit from this, but many can. The ideas of metric spaces and operators defined on them are important in applied mathematics and appear here in an elementary and natural way. Moreover, the contraction mapping theorem finds application in many parts of mathematics, and this seems to be a good place for the student to learn about it. To contrast this chapter with the previous ones, the approach here is analytical. Although everything up to this point pertained to initial value problems, a simple boundary value problem appears in this chapter as an application of the contraction mapping technique.

Chapter 4 treats linear boundary value problems, particularly the Sturm-Liouville problem, in one of the traditional ways — polar coordinate transformations. Ideas of inner products and orthogonality appear here in developing the rudiments of eigenfunction expansions. A nonlinear eigenvalue problem — a bifurcation problem — also appears, just to emphasize the effect of nonlinearities.

The book contains more material than can be covered in a semester. The instructor can pick and choose among the topics. Students in a course at this level will differ in ability and the material can be adjusted for this. I have usually taught Chapter 1 through the Putzer algorithm, skipped ahead and taught all of Chapter 2, presented the scalar existence theory in Chapter 3, and spent the remaining time in Chapter 4 (never completing it). Other routes through the material are possible, however, and the chapters are relatively independent. For example, Chapter 1 can be skipped entirely if students have a good background in systems (although a brief discussion of norms in R^n would help). I have made an effort to alert the reader when understanding of previous material is critical.

Professor John Baxley of Wake Forest University, Professor Juan Gatica of the University of Iowa, and Dr. Gail Wolkowicz of Emory University read the entire manuscript and made detailed comments. The presentation has benefited considerably from their many suggestions, and I wish to acknowledge their contributions and express my gratitude for their efforts. Several others — Gerald Armstrong of Brigham Young University, N. Cac of the University of Iowa, T. Y. Chow of California State University, Sacramento, Donald Smith of the University of California, San Diego, Joseph So of Emory University, and Monty Straus of Texas Tech University — read portions of the manuscript and made constructive comments. I gratefully acknowledge their assistance and express my appreciation for their efforts.

1

Systems of Linear Differential Equations

1. Introduction

Many problems in physics, biology, and engineering involve rates of change dependent on the interaction of the basic elements—particles, populations, charges, etc.—on each other. This interaction is frequently expressed as a system of ordinary differential equations, a system of the form

$$y'_1 = f_1(t, y_1, y_2, \ldots, y_n)$$
$$y'_2 = f_2(t, y_1, y_2, \ldots, y_n)$$
$$\vdots \qquad\qquad\qquad\qquad (1.1)$$
$$y'_n = f_n(t, y_1, y_2, \ldots, y_n).$$

Here the functions $f_i(t, y_1, \ldots, y_n)$ take values in R (the real numbers) and are defined on a set in R^{n+1} ($R \times R \times \cdots \times R$, $n + 1$ times). We seek a set of n unknown functions $(y_1(t), y_2(t), \ldots, y_n(t))$ defined on a real interval I such that, when these functions are inserted into the equations above, an identity results for every $t \in I$. In addition, certain other constraints (initial conditions or boundary conditions) may need to be satisfied. In this chapter we will be concerned with the special case that the functions f_j are linear in the variables y_i, $i = 1, \ldots,$

n. The problem takes the form

$$
\begin{aligned}
y_1' &= a_{11}(t)y_1 + a_{12}(t)y_2 + \cdots + a_{1n}(t)y_n + e_1(t) \\
y_2' &= a_{21}(t)y_1 + a_{22}(t)y_2 + \cdots + a_{2n}(t)y_n + e_2(t) \\
&\vdots \\
y_n' &= a_{n1}(t)y_1 + a_{n2}(t)y_2 + \cdots + a_{nn}(t)y_n + e_n(t).
\end{aligned}
\tag{1.2}
$$

In many applications the equations occur naturally in this form, or (1.2) may be an approximation to the nonlinear system (1.1). Moreover, some problems already familiar to the reader may be put into the form (1.2). For example, solving the second-order linear differential equation

$$
y'' + a(t)y' + b(t)y = e(t)
\tag{1.3}
$$

is equivalent to solving a system of the form

$$
\begin{aligned}
y_1' &= y_2 \\
y_2' &= -b(t)y_1 - a(t)y_2 + e(t).
\end{aligned}
\tag{1.4}
$$

To see the equivalence, suppose $(y_1(t), y_2(t))$ is a solution of (1.4). Then $y_1(t)$ is a solution of (1.3), since $y_2' = (y_1')' = -b(t)y_1 - a(t)y_1' + e(t)$, which is (1.3). On the other hand, if $y(t)$ is a solution of (1.3), then define $y_1(t) = y(t)$ and $y_2(t) = y'(t)$. This yields a solution of (1.4). Equation (1.3) is called a *scalar equation*; (1.4) is called a *system*.

The study of systems of the form (1.2) is made simpler by the use of matrix algebra. In the next section the basic notation, conventions, and theorems from linear algebra that are needed for the study of differential equations are collected for the convenience of the reader. Few proofs are given, and the reader meeting these concepts for the first time may wish to consult a textbook on linear algebra for an expanded development.

2. Some Elementary Matrix Algebra

If m and n are positive integers, an $m \times n$ *matrix* A is defined to be a set of mn numbers a_{ij}, $1 \leq i \leq m$, $1 \leq j \leq n$. (This is properly written as $a_{i,j}$ but the comma is omitted.) For notational purposes we write

$$A = \begin{bmatrix} a_{11} & a_{12} & \cdots & a_{1n} \\ a_{21} & & & a_{2n} \\ \vdots & & & \vdots \\ a_{m1} & & \cdots & a_{mn} \end{bmatrix} \quad \text{or} \quad A = \begin{pmatrix} a_{11} & a_{12} & \cdots & a_{1n} \\ a_{21} & & & a_{2n} \\ \vdots & & & \vdots \\ a_{m1} & & & a_{mn} \end{pmatrix};$$

that is, a_{ij} occupies a position in the ith row and the jth column of A. It is convenient to write

$$A = [a_{ij}]$$

to save space when the specific entries are not important or when they share a common property that can be illustrated by the bracket. For example,

$$A = \begin{bmatrix} \dfrac{2^i}{3^j} \end{bmatrix}, \qquad i = 1, 2, \quad j = 1, 2$$

denotes the matrix

$$A = \begin{bmatrix} \dfrac{2}{3} & \dfrac{2}{3^2} \\ \dfrac{2^2}{3} & \left(\dfrac{2}{3}\right)^2 \end{bmatrix}.$$

First, we will develop an algebra for these matrices. Then we will consider matrices with functions as entries and define continuous, differentiable, and integrable matrices.

Two $m \times n$ matrices $A = [a_{ij}]$, $B = [b_{ij}]$ are defined to be equal, written $A = B$, if $a_{ij} = b_{ij}$ for every i and j. Given $m \times n$ matrices A and B, we define their sum, $A + B$, by

$$A + B = [a_{ij} + b_{ij}].$$

For example, if

$$A = \begin{bmatrix} 1 & 0 & 1 \\ 2 & 2 & 7 \end{bmatrix}, \qquad B = \begin{bmatrix} 4 & -2 & -1 \\ 6 & 2 & -4 \end{bmatrix},$$

then

$$A + B = \begin{bmatrix} 1+4 & 0-2 & 1-1 \\ 2+6 & 2+2 & 7-4 \end{bmatrix} = \begin{bmatrix} 5 & -2 & 0 \\ 8 & 4 & 3 \end{bmatrix}.$$

From this definition of addition, it is obvious that if A, B, C are $m \times n$ matrices, then

$$A + B = B + A$$

and

$$A + (B + C) = (A + B) + C,$$

since numbers have these properties. We define multiplication of a matrix A by the number λ (called *scalar multiplication*) by

$$\lambda A = [\lambda a_{ij}].$$

Thus, for example, $-A = (-1)A = [-a_{ij}]$; or, if A is as in the example above and $\lambda = 2$, then

$$\lambda A = 2 \begin{bmatrix} 1 & 0 & 1 \\ 2 & 2 & 7 \end{bmatrix}$$

$$= \begin{bmatrix} 2 & 0 & 2 \\ 4 & 4 & 14 \end{bmatrix}.$$

If 0 denotes the matrix with all zero entries, that is, $a_{ij} = 0$, for all i and j, then

$$A - A = 0$$

and

$$A + 0 = A = 0 + A.$$

If A is an $m \times p$ matrix and B is a $p \times n$ matrix, the product of A and B, written AB, is defined to be the $m \times n$ matrix whose entries are given by

$$AB = \left[\sum_{k=1}^{p} a_{ik} b_{kj} \right]$$

$$= [c_{ij}], \qquad 1 \leq i \leq m, \quad 1 \leq j \leq n.$$

The ijth element of the product is the sum of the products of the elements in the ith row of A with the corresponding elements in the jth column of B. A simple example illustrates this definition.

$$A = \begin{bmatrix} 0 & 1 \\ 2 & 3 \end{bmatrix}, \qquad B = \begin{bmatrix} 4 & 5 & 6 \\ 7 & 8 & 9 \end{bmatrix}$$

$$AB = \begin{bmatrix} 0 \cdot 4 + 1 \cdot 7 & 0 \cdot 5 + 1 \cdot 8 & 0 \cdot 6 + 1 \cdot 9 \\ 2 \cdot 4 + 3 \cdot 7 & 2 \cdot 5 + 3 \cdot 8 & 2 \cdot 6 + 3 \cdot 9 \end{bmatrix}$$

$$= \begin{bmatrix} 7 & 8 & 9 \\ 29 & 34 & 39 \end{bmatrix}.$$

Note first that BA is not defined, since A is 2×2 and B is 2×3. The product is defined only when the number of columns of A is equal to the number of rows of B. If $n = m$, the matrix is said to be a *square matrix*. If A and B are square matrices of the same size, then both AB and BA are defined, but these need not be the same matrix. For example, if

$$A = \begin{bmatrix} 0 & 1 \\ 2 & 3 \end{bmatrix} \quad \text{and} \quad B = \begin{bmatrix} 2 & 3 \\ 4 & 5 \end{bmatrix},$$

then

$$AB = \begin{bmatrix} 0 & 1 \\ 2 & 3 \end{bmatrix}\begin{bmatrix} 2 & 3 \\ 4 & 5 \end{bmatrix} = \begin{bmatrix} 0 \cdot 1 + 1 \cdot 4 & 0 \cdot 3 + 1 \cdot 5 \\ 2 \cdot 2 + 3 \cdot 4 & 2 \cdot 3 + 3 \cdot 5 \end{bmatrix}$$

$$= \begin{bmatrix} 4 & 5 \\ 16 & 21 \end{bmatrix},$$

while

$$BA = \begin{bmatrix} 2 & 3 \\ 4 & 5 \end{bmatrix}\begin{bmatrix} 0 & 1 \\ 2 & 3 \end{bmatrix} = \begin{bmatrix} 2 \cdot 0 + 3 \cdot 2 & 2 \cdot 1 + 3 \cdot 3 \\ 4 \cdot 0 + 5 \cdot 2 & 4 \cdot 1 + 5 \cdot 3 \end{bmatrix}$$

$$= \begin{bmatrix} 6 & 11 \\ 10 & 19 \end{bmatrix},$$

and $AB \neq BA$.

The matrix $B = [b_{ij}]$, where $b_{ij} = a_{ji}$, is called the *transpose* of $A = [a_{ij}]$ and is denoted by A^T. If

$$A = \begin{bmatrix} 1 & 0 & 1 \\ 4 & 2 & 5 \\ -1 & -2 & 3 \end{bmatrix},$$

then

$$A^T = \begin{bmatrix} 1 & 4 & -1 \\ 0 & 2 & -2 \\ 1 & 5 & 3 \end{bmatrix}.$$

The rows and columns have been interchanged.

THEOREM 2.1

   The matrix product has the properties

    i. $A(BC) = (AB)C$

    ii. $\alpha(AB) = (\alpha A)B = A(\alpha B)$

    iii. $(A + B)C = AC + BC$

    iv. $C(A + B) = CA + CB$

    v. $(AB)^T = B^T A^T$

where A, B, C are matrices, α is a real or complex number, and the above products are defined.

The proofs of these properties are exercises in manipulating subscripts and are omitted.

A matrix with $n = 1$ (i.e., an $n \times 1$ matrix)

$$\begin{bmatrix} a_{11} \\ a_{21} \\ \vdots \\ a_{m1} \end{bmatrix}$$

is called an *n*-dimensional (*column*) *vector*. The matrix I (denoted I_p if the dimension $p \times p$ is important), called the *identity matrix*, is defined by

$$I = \begin{bmatrix} 1 & 0 & & \cdots & 0 \\ 0 & 1 & & \cdots & 0 \\ 0 & 0 & 1 & \cdots & 0 \\ \vdots & & & \ddots & \\ 0 & 0 & 0 & \cdots & 1 \end{bmatrix},$$

that is, $a_{ii} = 1$ and $a_{ij} = 0$, $i \neq j$. If A is an $m \times n$ matrix and I_n is the $n \times n$ identity matrix, then, from the definition of multiplication, it follows that

$$AI_n = A.$$

If I_m is the $m \times m$ identity matrix,

$$I_m A = A.$$

Our interest is principally in vectors and square matrices.

Let A be an $n \times n$ matrix. A real number called the *determinant* of A is associated with each square matrix. The definition of this real number is inductive on n. If $n = 1$, $\det A = a_{11}$. Suppose that $\det A$ has been defined for $n = k \geq 1$. Given an element a_{ij} of a matrix A, M_{ij}, the *minor* of a_{ij}, is the matrix obtained from A by deleting the ith row and the jth column. A_{ij}, the *cofactor* of a_{ij}, is defined by

$$A_{ij} = (-1)^{i+j} \det M_{ij}.$$

For $n = k + 1$, we define

$$\det A = a_{11} A_{11} + a_{21} A_{21} + \cdots + a_{n1} A_{n1}.$$

The following examples clarify this definition. If

$$A = \begin{bmatrix} a_{11} & a_{12} \\ a_{21} & a_{22} \end{bmatrix},$$

then $\det A = a_{11} a_{22} - a_{21} a_{12}$. If

$$A = \begin{bmatrix} a_{11} & a_{12} & a_{13} \\ a_{21} & a_{22} & a_{23} \\ a_{31} & a_{32} & a_{33} \end{bmatrix},$$

then

$$\det A = a_{11} \det \begin{bmatrix} a_{22} & a_{23} \\ a_{32} & a_{33} \end{bmatrix} - a_{21} \det \begin{bmatrix} a_{12} & a_{13} \\ a_{32} & a_{33} \end{bmatrix} + a_{31} \det \begin{bmatrix} a_{12} & a_{13} \\ a_{22} & a_{23} \end{bmatrix}$$

$$= a_{11} a_{22} a_{33} - a_{11} a_{23} a_{32} - a_{21} a_{12} a_{33} + a_{21} a_{13} a_{32}$$

$$+ a_{31} a_{12} a_{23} - a_{31} a_{13} a_{22}.$$

By definition, a determinant, for $n > 1$, is the sum of all of the elements of the first column multiplied by their cofactors. Actually, the first column need not necessarily be used to find a determinant; in fact, the following is also true.

THEOREM 2.2

$$\det A = \sum_{i=1}^{n} a_{ij} A_{ij} = \sum_{j=1}^{n} a_{ij} A_{ij}.$$

The content of this theorem is that in the preceding inductive definition of a determinant, the first column can be replaced by an arbitrary column or an arbitrary row. We will accept this theorem without proof. An important property of expansions of the type in the theorem is that

$$\sum_{i=1}^{n} a_{ij} A_{ik} = \sum_{j=1}^{n} a_{ij} A_{kj} = 0,$$

if $j \neq k$ on the left and $i \neq k$ on the right. That is, if the cofactors are taken from a different column or a different row, the resulting sum is zero.

An important property of determinants (which also will not be proved here) is given in the following.

THEOREM 2.3

If A and B are $n \times n$ matrices,

$$\det (AB) = (\det A)(\det B);$$

that is, the determinant of the product of two matrices is the product of the determinants.

A square matrix is said to be *singular* if $\det A = 0$ and *nonsingular* if $\det A \neq 0$.

A matrix B is called the *inverse* of the square matrix A if $AB = I$. Suppose $\det A \neq 0$. We define B by

$$B = \frac{1}{\det A} [A_{ij}]^{T}; \qquad (2.1)$$

that is, B is the transpose of the matrix obtained by replacing each element by its cofactor and then dividing by the scalar $\det A$. Then

$$AB = \frac{1}{\det A} \left[\sum_{k=1}^{n} a_{ik} A_{jk} \right].$$

If $i = j$, then the entry is $\det A$; if $i \neq j$, the product is zero, as noted previously.

Hence, the matrix $[\sum_{k=1}^{n} a_{ik} A_{jk}]$ looks like

$$
\begin{bmatrix}
\det A & 0 & 0 & \cdots & 0 \\
0 & \det A & 0 & \cdots & \\
\vdots & & & & \\
0 & & \cdots & & \det A
\end{bmatrix}.
$$

Therefore, $AB = I$, so B is the inverse of A. This matrix B is usually written A^{-1}. Since $AA^{-1} = I$, then $1 = \det(AA^{-1}) = (\det A)(\det A^{-1})$, and it cannot be the case that A has an inverse if $\det A = 0$. The arguments above can be used to show the following.

THEOREM 2.4
A necessary and sufficient condition that A^{-1} exists is that $\det A \neq 0$.

It can also be shown that $A^{-1}A = I$, that is, that A and A^{-1} commute. Further, A^{-1} is unique. To see this, suppose there exists a matrix X such that $AX = I$. Then, multiplying both sides of this equation on the left by A^{-1} yields

$$A^{-1}AX = A^{-1}I = A^{-1}$$

or

$$X = A^{-1}.$$

Similarly, if $XA = I$, then $XAA^{-1} = A^{-1}$, or $X = A^{-1}$.
If A_1 and A_2 are nonsingular, then

$$(A_1 A_2)^{-1} = A_2^{-1} A_1^{-1},$$

since $A_1 A_2 A_2^{-1} A_1^{-1} = I$.
To fit initial conditions for solutions of systems of differential equations, it will be necessary to consider systems of linear algebraic equations in the variables $x_1, \ldots, x_n$ of the form

$$a_{11}x_1 + a_{12}x_2 + \cdots + a_{1n}x_n = c_1$$
$$a_{21}x_1 + \quad \cdots \quad + a_{2n}x_n = c_2$$
$$\vdots$$
$$a_{n1}x_1 + \quad \cdots \quad + a_{nn}x_n = c_n.$$

This can be written

$$Ax = c, \tag{2.2}$$

where $A = [a_{ij}]$ is an $n \times n$ matrix and $x = [x_i]$ and $c = [c_i]$ are vectors. If $\det A \neq 0$, then

$$x = A^{-1}c$$

is a solution, since

$$Ax = A(A^{-1}c) = AA^{-1}c = c.$$

It is not difficult to see that this is the only solution. Suppose x and y are both solution vectors of (2.2). Since $Ax = c$ and $Ay = c$, $Ax - Ay = c - c = 0$, the null vector (all entries zero). The distributive law (x and y are $n \times 1$ matrices) says that

$$A(x - y) = 0.$$

Since A is invertible and $A^{-1}0 = 0$, we have

$$A^{-1}A(x - y) = A^{-1}0 = 0;$$

hence

$$x - y = 0$$

or

$$x = y.$$

Thus there is only one solution. The converse of this result is also true.

THEOREM 2.5

A necessary and sufficient condition for the system (2.2) to have a unique solution is that A be nonsingular.

In particular, note that if the vector c is null, that is, has all its components zero, $x = 0$ is the only solution vector if A is nonsingular. (If A is singular, then $x = 0$ is one solution and there *must* be another solution, since solutions are *not* unique.)

A finite set of n-dimensional vectors $x_1, \ldots, x_k$, that is, $n \times 1$ matrices, is said to be *linearly dependent* if there exist constants $c_1, \ldots, c_k$, not all zero, such that

$$c_1 x_1 + c_2 x_2 + \cdots + c_k x_k = 0.$$

A set of vectors that is not linearly dependent is said to be *linearly independent*. An expression of the form $c_1 x_1 + c_2 x_2 + \cdots + c_k x_k$ is said to be a *linear combination* of the vectors $x_1, \ldots, x_n$. The following theorem offers a way to check whether a matrix is singular.

THEOREM 2.6

A necessary and sufficient condition for a matrix to be nonsingular is that its columns are linearly independent vectors.

Proof. Let A be a matrix and let $x^1, \ldots, x^n$ denote the n-dimensional column vectors of A. We inquire whether there exist real numbers $c_1, \ldots, c_n$ such that

$$c_1 x^1 + c_2 x^2 + \cdots + c_n x^n = 0. \tag{2.3}$$

If we let C be the vector

$$C = \begin{bmatrix} c_1 \\ c_2 \\ \vdots \\ c_n \end{bmatrix},$$

then (2.3) may be written

$$AC = 0, \tag{2.4}$$

since $A = [x^1, \ldots, x^n]$. The equation $AC = 0$ has a nontrivial solution (Theorem 2.5) if and only if A is singular. Thus, if A is nonsingular, the only solution of (2.3) is $c_1 = c_2 = \cdots = c_n = 0$ and the vectors $x^1, \ldots, x^n$ are linearly independent. If A is singular, there exists a nontrivial solution C of (2.4) and $x^1, \ldots, x^n$ are linearly dependent, using the components of C as the constants in the definition of linear dependence.

Finally, it will be necessary to consider matrices whose elements are functions. We can think of a matrix $A(t)$ as a mapping from the set of real numbers into the set of $n \times n$ matrices. It is simpler, however, to think of such a matrix as n^2 functions labeled $a_{ij}(t)$ and make our definitions in terms of the entries rather than in terms of the mapping. Proceeding this way, a matrix of functions $A(t) = [a_{ij}(t)]$ is said to be

1. *continuous* at a point t_0 if each $a_{ij}(t)$ is continuous at t_0,

2. *differentiable* at a point t_0 if each $a_{ij}(t)$ is differentiable at t_0,

3. *integrable* over $[a, b]$ if each $a_{ij}(t)$ is integrable over $[a, b]$,

If $A(t)$ is differentiable, define $A'(t) = [a'_{ij}(t)]$. From the definition of the product of two matrices, we have at once that if $A(t)$ and $B(t)$ are differentiable and the product $A(t)B(t)$ is defined, then $A(t)B(t)$ is differentiable. Further,

$$(AB)' = A'B + AB'. \tag{2.5}$$

To see this, note that

$$(AB)'(t) = \left[\sum_k (a_{ik}(t)b_{kj}(t))' \right]$$

$$= \left[\sum_k a'_{ik}(t)b_{kj}(t) + a_{ik}(t)b'_{kj}(t) \right]$$

$$= \left[\sum_k a'_{ik}(t)b_{kj}(t) \right] + \left[\sum_k a_{ik}(t)b'_{kj}(t) \right]$$

$$= A'B + AB'.$$

This fact will be very important in our development of a theory for systems of differential equations. Note that the order of multiplication is important.

Similarly, we define

$$\int_0^t A(t)\,ds = \left[\int_0^t a_{ij}(s)\,ds \right],$$

and the usual rules for integration apply. For example,

$$\int_0^t [A(s) + B(s)]\,ds = \int_0^t A(s)\,ds + \int_0^t B(s)\,ds.$$

EXERCISES

1. Find $A + B$, AB, and BA when

   (a) $A = \begin{bmatrix} 1 & 2 \\ 3 & 4 \end{bmatrix}$, $B = \begin{bmatrix} 1 & 0 \\ -1 & 2 \end{bmatrix}$

(b) $A = \begin{bmatrix} 3 & 2 \\ -1 & 4 \end{bmatrix}$, $B = \begin{bmatrix} 7 & -1 \\ 4 & 2 \end{bmatrix}$

(c) $A = \begin{bmatrix} 1 & 2 & 3 \\ 0 & 1 & 0 \\ 4 & 5 & 6 \end{bmatrix}$, $B = \begin{bmatrix} 2 & 0 & 1 \\ 0 & -1 & 4 \\ 2 & 2 & 0 \end{bmatrix}$

(d) $A = \begin{bmatrix} -1 & -2 & -3 \\ 1 & 2 & 3 \\ 0 & 1 & -4 \end{bmatrix}$, $B = \begin{bmatrix} 2 & 1 & 3 \\ -1 & 4 & 2 \\ 6 & 1 & 0 \end{bmatrix}$

(e) $A = \begin{bmatrix} 1 & 1 & 0 & 0 \\ 0 & 1 & 1 & 0 \\ 0 & 0 & 1 & 0 \\ 0 & 0 & 0 & 2 \end{bmatrix}$, $B = \begin{bmatrix} 2 & 0 & 0 & 0 \\ 0 & 3 & 1 & 0 \\ 0 & 0 & 3 & 0 \\ 0 & 0 & 0 & 1 \end{bmatrix}$

2. A matrix is said to be *diagonal* if $a_{ij} = 0$ when $i \neq j$. Show that the product of two diagonal matrices is diagonal.

3. Show that $A(BC) = (AB)C$ by using the definition of product for matrices.

4. If α is a scalar and A and B are matrices, show that $\alpha(AB) = (\alpha A)B = A(\alpha B)$, that is, scalars (numbers) "factor through" matrix multiplication.

5. Establish the distributive laws for matrix multiplication,

$$(A + B)C = AC + BC$$

$$C(A + B) = CA + CB.$$

6. Prove that $(AB)^T = B^T A^T$.

7. Construct A^{-1}, if $A =$

(a) $\begin{bmatrix} 1 & 2 \\ 3 & 4 \end{bmatrix}$ **(b)** $\begin{bmatrix} 2 & 0 & 0 \\ 0 & 3 & 0 \\ 0 & 0 & 4 \end{bmatrix}$

(c) $\begin{bmatrix} 2 & 1 & 0 \\ 0 & 2 & 0 \\ 0 & 0 & 3 \end{bmatrix}$ **(d)** $\begin{bmatrix} 1 & 0 & 1 \\ 0 & 1 & 0 \\ 1 & 1 & 0 \end{bmatrix}$

(e) $\begin{bmatrix} 2 & 1 & 0 & 0 \\ 0 & 2 & 1 & 0 \\ 0 & 0 & 2 & 0 \\ 0 & 0 & 0 & 2 \end{bmatrix}$

8. Find solutions of the system

$$Ax = \begin{pmatrix} 1 \\ 0 \\ 1 \end{pmatrix}$$

when A is as in Exercise 7(b), (c), and (d).

9. Determine whether the following sets of vectors are linearly independent or linearly dependent.

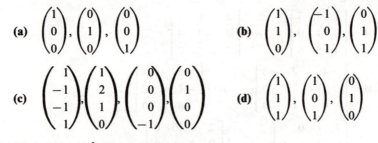

(a) $\begin{pmatrix} 1 \\ 0 \\ 0 \end{pmatrix}, \begin{pmatrix} 0 \\ 1 \\ 0 \end{pmatrix}, \begin{pmatrix} 0 \\ 0 \\ 1 \end{pmatrix}$ (b) $\begin{pmatrix} 1 \\ 1 \\ 0 \end{pmatrix}, \begin{pmatrix} -1 \\ 0 \\ 1 \end{pmatrix}, \begin{pmatrix} 0 \\ 1 \\ 1 \end{pmatrix}$

(c) $\begin{pmatrix} 1 \\ -1 \\ -1 \\ 1 \end{pmatrix}, \begin{pmatrix} 1 \\ 2 \\ 1 \\ 0 \end{pmatrix}, \begin{pmatrix} 0 \\ 0 \\ 0 \\ -1 \end{pmatrix}, \begin{pmatrix} 0 \\ 1 \\ 0 \\ 0 \end{pmatrix}$ (d) $\begin{pmatrix} 1 \\ 1 \\ 1 \end{pmatrix}, \begin{pmatrix} 1 \\ 0 \\ 1 \end{pmatrix}, \begin{pmatrix} 0 \\ 1 \\ 0 \end{pmatrix}$

10. Find $A'(t)$ and $\int_0^t A(s)\,ds$ if $A(t) =$

(a) $\begin{pmatrix} \cos t & \sin t \\ -\sin t & \cos t \end{pmatrix}$ (b) $\begin{pmatrix} t & t^2 \\ 1 & t \end{pmatrix}$

(c) $\begin{pmatrix} t & t^3 \\ t^2 & t^4 \end{pmatrix}$ (d) $\begin{pmatrix} e^t & e^t \sin t \\ e^t & e^t \cos t \end{pmatrix}$

11. Verify equation (2.5) where $A(t)$ is given by Exercise 10(a) and $B(t)$ by Exercise 10(b).

3.   The Structure of Solutions of Homogeneous Linear Systems

Let x represent an n-dimensional vector, let A be an $n \times n$ matrix of continuous functions defined on an interval I, and let $e(t)$ be an n-vector of continuous functions defined on I. The system (1.2), using matrix notation, can be written

$$x' = A(t)x + e(t). \tag{3.1}$$

A solution of (3.1) on I is a differentiable vector function $\varphi(t)$ such that

$$\varphi'(t) = A(t)\varphi(t) + e(t)$$

for every $t \in I$. If x_0 is a constant vector and $t_0 \in I$, then the initial value problem for (3.1) is to find a solution of (3.1) that satisfies, in addition,

$$\varphi(t_0) = x_0. \tag{3.2}$$

We state the basic existence theorem, which is a special case of a theorem to be proved in Chapter 3.

THEOREM 3.1

Let $A(t)$ be a continuous $n \times n$ matrix defined on an interval I and let $e(t)$ be a continuous n-vector defined on I. For every constant n-vector x_0 and every $t_0 \in I$, there exists a unique differentiable vector $\varphi(t)$ defined on I such that

$$\varphi'(t) = A(t)\varphi(t) + e(t), \qquad t \in I,$$

$$\varphi(t_0) = x_0.$$

For the remainder of this section we consider the case $e(t) = 0$, called the *homogeneous case*. (The quantity $e(t)$ sometimes represents an external force, so this case is also called the *unforced case*.) Here we attempt to develop a structure similar to that for scalar equations. It will be convenient to think of Equation (3.1) in a slightly different way. Let $\mathscr{A}$ be one set of functions and $\mathscr{B}$ another. Suppose that for each element x in the set $\mathscr{A}$ we associate a unique element of the set $\mathscr{B}$, called Tx. T is a mapping from the set $\mathscr{A}$ into the set $\mathscr{B}$ and symbolically we write $T: \mathscr{A} \to \mathscr{B}$. The mapping T is also called an *operator*. $\mathscr{A}$ is called the *domain* of T and $\mathscr{B}$, the set of all y such that $y = Tx$, is called the *range* of T. (Sometimes it is convenient to indicate a larger set, a set containing the range, in the symbolic definition.) For example, let $\mathscr{A}$ be the set of all continuous n-vectors on $[0, 1]$. Define $Tx = y$, $x \in \mathscr{A}$, by

$$y(t) = \int_0^t x(s)\, ds.$$

Set $\mathscr{B}$, in this case, could be the set of continuous vectors defined on $[0, 1]$ that have a continuous derivative on $(0, 1)$. As another example, let $\mathscr{A}$ be as above and let Ω be an $n \times n$ constant matrix; then an operator can be defined by $y = \Omega x$. All of our sets will have the property that if α is a number and x and y are in the set, αx and $x + y$ are in the set.

Now let $x(t)$ be a continuously differentiable vector. Define an operator L on the set of all such functions by

$$L[x] = x' - Ax,$$

where A is an $n \times n$ continuous matrix. L maps continuously differentiable functions into continuous ones and a solution of (3.1) is just the set of functions that are mapped by L onto the function $e(t)$, or, in the homogeneous case, onto the constant function that is everywhere zero.

An operator T is said to be *linear* if for any two elements x, y in its domain, and any two numbers (scalars) α and β,

$$T(\alpha x + \beta y) = \alpha T(x) + \beta T(y).$$

THEOREM 3.2
The operator L is a linear operator.

Proof. Let $x_1(t)$, $x_2(t)$ be differentiable vector functions and let c_1, c_2 be real constants. Then

$$L[c_1 x_1 + c_2 x_2](t) = (c_1 x_1(t) + c_2 x_2(t))' - A(t)(c_1 x_1(t) + c_2 x_2(t))$$
$$= c_1 x_1'(t) + c_2 x_2'(t) - c_1 A(t) x_1(t) - c_2 A(t) x_2(t)$$
$$= c_1 (x_1'(t) - A(t) x_1(t)) + c_2 (x_2'(t) - A(t) x_2(t))$$
$$= c_1 L[x_1](t) + c_2 L[x_2](t).$$

THEOREM 3.3
   Every linear combination of solutions of

$$L[x] = 0 \tag{3.3}$$

   is a solution of (3.3).

Proof. If $x_1(t)$, $x_2(t)$ are solutions of (3.3) and c_1 and c_2 are constants,

$$L[c_1 x_1 + c_2 x_2] = c_1 L[x_1] + c_2 L[x_2] = 0,$$

since $L[x_1] = 0$ and $L[x_2] = 0$.

Suppose now that we have n solution vectors, $x_i(t)$, of (3.3) defined on an interval I. Then we can form a matrix Φ whose columns are these solutions. We write

$$\Phi = [x_1, x_2, \ldots, x_n].$$

Since the elements of Φ are differentiable, we can compute Φ'. Now

ith column $\Phi'(t) = x_i'(t) = A(t) x_i(t)$,

so that

$$\Phi'(t) = [A(t) x_1(t), A(t) x_2(t), \ldots, A(t) x_n(t)]$$
$$= A(t)[x_1(t), x_2(t), \ldots, x_n(t)]$$
$$= A(t)\Phi(t).$$

That is, Φ satisfies

$$\Phi'(t) = A(t)\Phi(t). \tag{3.4}$$

Equation (3.4) is a shorthand method for writing n-vector differential equations (n^2 scalar differential equations). For this reason, Theorem 3.1—the existence and uniqueness theorem—applies if we specify an initial condition $\Phi(0) = C$, where C is a constant matrix. Using (3.4), we have the following useful fact.

If Φ *is a matrix whose columns are solutions of (3.3) and* c *is a constant vector, then* Φc *is a solution of (3.3).*

Proof. The proof is a straightforward computation,

$$L[\Phi c] = (\Phi c)' - A\Phi c$$
$$= \Phi' c - A\Phi c$$
$$= (\Phi' - A\Phi)c$$
$$= 0,$$

since Φ satisfies (3.4).

If $\Phi(t)$ is a matrix that is nonsingular for each t and that satisfies the matrix differential equation (3.4), then Φ is said to be a *fundamental matrix* for the differential equation $x' - Ax = 0$.

We are now equipped to prove the principal theorem of this section. The content of this theorem is that finding any fundamental matrix for (3.3) allows us to find all of the solutions of (3.3).

THEOREM 3.4

If Φ **is a fundamental matrix for (3.3) on an interval** I **where** $A(t)$ **is continuous, then every solution of (3.3) can be written** Φc **for an appropriate constant vector** c.

Proof. Let $x(t)$ be an arbitrary solution defined on an interval I, let $t_0 \in I$, and let Φ be a given fundamental matrix for (3.3). Now $\Phi(t_0)$ is a nonsingular constant matrix, so $\Phi^{-1}(t_0)$ exists (Theorem 2.4). Let $c = \Phi^{-1}(t_0)x(t_0)$. Since c is a vector, $y(t) = \Phi(t)c$ is a solution of (3.3).

Furthermore,

$$y(t_0) = \Phi(t_0)c = \Phi(t_0)\Phi^{-1}(t_0)x(t_0)$$
$$= x(t_0).$$

Since solutions of the initial value problem are unique (Theorem 3.1), it follows that $x(t) = y(t)$ for $t \in I$, or that $x(t) \equiv \Phi(t)c$, as claimed.

The reader will recognize that Theorem 3.4 is a wholesale way of doing what was done in particular for scalar equations. For example, an arbitrary solution of a second-order scalar equation can be represented as a linear combination of two linearly independent solutions. The quantity $y(t) = \Phi(t)c$ represents the ith component of $y(t)$ as a linear combination of the ith components of n given solutions—each row is in fact the same linear combination, and the components of c are the coefficients.

We illustrate Theorem 3.4 by comparing it with the theory for second-order scalar equations. Consider

$$
\begin{aligned}
y_1' &= y_2 \\
y_2' &= -y_1.
\end{aligned}
\tag{3.5}
$$

This system is the system we obtain from

$$y'' + y = 0. \tag{3.6}$$

By substitution in (3.5) it may be verified that the vectors $\begin{bmatrix} \sin(t) \\ \cos(t) \end{bmatrix}$ and $\begin{bmatrix} \cos(t) \\ -\sin(t) \end{bmatrix}$ are two solutions of (3.5). Let

$$
\Phi = \begin{bmatrix} \sin(t) & \cos(t) \\ \cos(t) & -\sin(t) \end{bmatrix}.
$$

Since $\det \Phi = -\sin^2(t) - \cos^2(t) = -1$, Φ is nonsingular and hence is a fundamental matrix. By Theorem 3.4, any solution $y(t)$ of (3.5) can be written

$$
y(t) = \Phi c = \begin{bmatrix} \sin(t) & \cos(t) \\ \cos(t) & -\sin(t) \end{bmatrix} \begin{bmatrix} c_1 \\ c_2 \end{bmatrix}
$$

for appropriate c_1 and c_2.

Note that two linearly independent solutions of (3.6) are $\varphi_1 = \sin(t)$, $\varphi_2 = \cos(t)$. *The Wronskian of φ_1 and φ_2, $W(\varphi_1, \varphi_2)(t)$, is the determinant of the fundamental matrix φ.* That is,

$$
\det \Phi(t_0) = W(\varphi_1, \varphi_2)(t).
$$

This is the connecting link between the theory for second-order scalar equations and the theory for the equivalent systems.

Consider now the system ($n = 3$),

$$y_1' = y_2 + 4y_3$$
$$y_2' = -y_1 - 2y_3$$
$$y_3' = y_3,$$

or

$$\begin{pmatrix} y_1 \\ y_2 \\ y_3 \end{pmatrix}' = \begin{pmatrix} 0 & 1 & 4 \\ -1 & 0 & -2 \\ 0 & 0 & 1 \end{pmatrix} \begin{pmatrix} y_1 \\ y_2 \\ y_3 \end{pmatrix}. \qquad (3.7)$$

It can be verified by substitution that three solutions are

$$\begin{bmatrix} \sin(t) \\ \cos(t) \\ 0 \end{bmatrix}, \quad \begin{bmatrix} \cos(t) \\ -\sin(t) \\ 0 \end{bmatrix}, \quad \text{and} \quad \begin{bmatrix} e^t \\ -3e^t \\ e^t \end{bmatrix}.$$

Then Φ is given by

$$\Phi(t) = \begin{bmatrix} \sin(t) & \cos(t) & e^t \\ \cos(t) & -\sin(t) & -3e^t \\ 0 & 0 & e^t \end{bmatrix}. \qquad (3.8)$$

A computation shows that

$$\det \Phi(t) = e^t(-\sin^2(t) - \cos^2(t))$$
$$= -e^t \neq 0,$$

so Φ is nonsingular and hence is a fundamental matrix. Every solution of (3.7) can be written $\Phi(t)c$ for an appropriate constant vector c.

Suppose, for example, we desire to find the solution of (3.7) that satisfies the initial conditions $y_1(0) = 1$, $y_2(0) = 1$, $y_3(0) = 1$. It is necessary to choose

$$c = \begin{bmatrix} c_1 \\ c_2 \\ c_3 \end{bmatrix}$$

such that

$$\Phi(0)c = \begin{bmatrix} 1 \\ 1 \\ 1 \end{bmatrix}$$

or

$$\begin{bmatrix} 0 & 1 & 1 \\ 1 & 0 & -3 \\ 0 & 0 & 1 \end{bmatrix} \begin{bmatrix} c_1 \\ c_2 \\ c_3 \end{bmatrix} = \begin{bmatrix} 1 \\ 1 \\ 1 \end{bmatrix}.$$

In equation form, this is

$$c_2 + c_3 = 1$$
$$c_1 - 3c_3 = 1$$
$$c_3 = 1.$$

(We could, of course, invert the matrix, but that would involve more labor.) Thus

$$c = \begin{bmatrix} 4 \\ 0 \\ 1 \end{bmatrix},$$

and

$$\Phi(t)c = \begin{bmatrix} \sin(t) & \cos(t) & e^t \\ \cos(t) & -\sin(t) & -3e^t \\ 0 & 0 & e^t \end{bmatrix} \begin{bmatrix} 4 \\ 0 \\ 1 \end{bmatrix}$$

$$= \begin{bmatrix} 4\sin(t) + e^t \\ 4\cos(t) - 3e^t \\ e^t \end{bmatrix},$$

is the desired solution.

The principal issue, however, the question of the existence of a fundamental matrix (and how to find it), has been sidestepped. In the examples above the matrix was exhibited explicitly, but the question as to whether a fundamental matrix can always be found thus far remains open. Let

$$\varphi^i(t) = \begin{bmatrix} \varphi_1^i(t) \\ \vdots \\ \varphi_n^i(t) \end{bmatrix}$$

be the solution of (3.3) satisfying $\varphi_j^i(t_0) = 0, j \neq i$, and $\varphi_i^i(t_0) = 1$, that is,

$$\varphi^i(t_0) = \begin{bmatrix} 0 \\ \vdots \\ 1 \\ 0 \\ \vdots \\ 0 \end{bmatrix} \leftarrow i\text{th place.}$$

The set of solutions $\varphi^i(t)$, $i = 1, 2, \ldots, n$, which exists by Theorem 3.1, can be used to form a solution matrix

$$\Phi(t) = [\varphi^1(t), \varphi^2(t), \ldots, \varphi^n(t)].$$

Further, for $t = 0$, $\det \Phi(t) = 1$. If the determinant should remain nonzero on an interval I, then Φ would be a fundamental matrix on this interval. This is indeed the case, as given by Theorem 3.5. The *trace* of a matrix $A(t) = [a_{ij}(t)]$ (written $\operatorname{tr} A(t)$) is defined to be the sum of the diagonal elements, that is, $\operatorname{tr} A(t) = \sum_{i=1}^n a_{ii}(t)$. Note that $\operatorname{tr} A(t)$ is a scalar function.

THEOREM 3.5   *(Abel's formula)*

Let $A(t)$ be an $n \times n$ matrix of continuous functions on $I = [a, b]$ and let $\Phi(t)$ be a matrix of differentiable functions such that

$$\Phi'(t) = A(t)\Phi(t).$$

Then for $t, t_0 \in I$,

$$\det \Phi(t) = \det [\Phi(t_0)] e^{\int_{t_0}^t \operatorname{tr} A(s)\, ds}.$$

Since an exponential is never zero (note that this exponential is a real-valued function, not a matrix), Theorem 3.5 says that if Φ is a matrix whose columns are solutions of (3.3), then $\det \Phi(t)$ is, identically zero or never zero. Thus, to find a fundamental matrix, we need only to find n solutions that at some point t_0 are linearly independent vectors. We omit the proof of Theorem 3.5, although the exercises following this section indicate how it can be done. Combining the preceding arguments, we have

---

THEOREM 3.6

If $A(t)$ is continuous, there exists a fundamental matrix for the system (3.3).

EXERCISES

1. If $\Phi(t)$ is a fundamental matrix for $x' = Ax$ and C is a nonsingular matrix of the same dimension, show that $\Phi(t)C$ is a fundamental matrix. (Recall that $C' = 0$, if C is a constant matrix.)

2. Show that if $\Phi(t)$ and $\Psi(t)$ are fundamental matrices for $x' = Ax$, then there is a constant, nonsingular matrix C such that $\Phi(t)C = \Psi(t)$.

3. Verify that the matrix (3.8) is a fundamental matrix for (3.7). Illustrate Theorem 3.5 with the matrix (3.8).

4. Define an operator Tx on the set of continuous functions on $[0, 1]$ by $(Tx)(t) = \int_0^t x(s)\, ds$, $t \in [0, 1]$. Show that T is a linear operator. Can you define T on a larger domain?

5. Show that multiplication of vectors in R^n by a (fixed) matrix A defines a linear operator.

6. Let $B(t) = \begin{bmatrix} b_{11}(t) & b_{12}(t) \\ b_{21}(t) & b_{22}(t) \end{bmatrix}$. Let $b_{ij}(t)$ be differentiable. Compute $(\det B(t))'$ by first expanding $\det B(t)$ and then differentiating. Then show that

$$(\det(B(t)))' = \det\begin{bmatrix} b'_{11}(t) & b'_{12}(t) \\ b_{21}(t) & b_{22}(t) \end{bmatrix} + \det\begin{bmatrix} b_{11}(t) & b_{12}(t) \\ b'_{21}(t) & b'_{22}(t) \end{bmatrix}.$$

7. Let $\Phi(t) = \begin{bmatrix} x_{11}(t) & x_{12}(t) \\ x_{21}(t) & x_{22}(t) \end{bmatrix}$ be a fundamental matrix for $x' = A(t)$, where $A = \begin{bmatrix} a_{11}(t) & a_{12}(t) \\ a_{21}(t) & a_{22}(t) \end{bmatrix}$. Show that

$$(\det \Phi)' = \left(\det\begin{bmatrix} \sum_1^2 a_{1k}x_{k1} & \sum_1^2 a_{1k}x_{k2} \\ x_{21} & x_{22} \end{bmatrix} + \det\begin{bmatrix} x_{11} & x_{12} \\ \sum_1^2 a_{2k}x_{k1} & \sum_1^2 a_{2k}x_{k2} \end{bmatrix} \right)$$

$$= \sum_1^2 a_{ii} \det \Phi.$$

8. Let $z(t) = \det \Phi(t)$ where $\Phi(t)$ is as given in Exercise 7. Use Exercise 7 to conclude that $z(t) = z(0)e^{\int_0^t (\operatorname{tr} A(s))\, ds}$ and establish Theorem 3.5 for this special case.

4.   Matrix Analysis and the Matrix Exponential

In Section 2 some of the basic ideas of matrices and their algebra were developed. Matrices were used in Section 3, but mostly for notational convenience; they were used to give a simple representation to complicated expressions. We now need to take a further step, to learn how to take a limit of a sequence of matrices. The power of the simple idea of limit is familiar to every calculus student and lies at the heart of all of analysis. It is possible to carry over to matrices (and to more general settings) many of the ideas from elementary calculus. We limit our scope to one simple idea, convergence. We will use this notion to define a very useful matrix, which is a sum of an infinite series of matrices. This matrix, called the *exponential* of another matrix, has many (and fails to have many other) properties of the real exponential function. It is an example of one of the fundamental themes of mathematics, taking an idea in one setting and developing it in another—the process called generalization. As we shall see in the material that follows, the concept of the exponential of a matrix is a very useful generalization.

The reader is assumed to be familiar with limits of sequences and sums of series from calculus. The notion of making the absolute value (or modulus, in the complex case) of a quantity small is crucial. The first step on the way to the definition we need is to replace absolute value with a notion applicable to vectors and matrices. This concept is called the *norm* of a matrix or a vector. There are many ways to do this, and for some applications the clever choice of a norm is very important. For what we need here, any of the usual notions of norm would be satisfactory, so we choose one that makes the proofs easy.

Define, for an $r \times r$ matrix A, a real number, called the norm of A, written $\|A\|$, by

$$\|A\| = \sum_{i,j} |a_{ij}|.$$

If A is a vector $(a_1, \ldots, a_r)^T$, define $\|A\| = \sum_{i=1}^{r} |a_i|$. The norm of a matrix has the following properties:

1. $\|A\| > 0, A \neq 0$, and $\|0\| = 0$;

2. $\|cA\| = |c| \|A\|$;

3. $\|A + B\| \leq \|A\| + \|B\|$; and

4. $\|AB\| \leq \|A\| \|B\|$,

where A and B are $r \times r$ matrices and c is a real or complex number.

If x is an n-vector, the definitions for norms of matrices and vectors are so related that $\|Ax\| \leq \|A\| \|x\|$. Note that the norm of a vector satisfies properties (1), (2), and (3). We also note that there are other "norms" for vectors and matrices.

The definitions that follow distinguish between the concept of convergence and the concept of converging to a limit. The distinction is not really needed at this point, but it will be important in Chapter 3 where these ideas must be extended further.

A sequence of $r \times r$ matrices, A_n, is *convergent* (or is a *Cauchy sequence*) if for $\varepsilon > 0$ there exists a positive integer N such that if $m, n > N$, then $\| A_n - A_m \| < \varepsilon$. The definition of convergence of a sequence of matrices is exactly the same as the definition of convergence of a sequence of real numbers except that norm, $\| \quad \|$, has replaced absolute value, $| \quad |$. This is true of the definition of limit of a sequence of matrices as well. A matrix A is said to be the *limit* of a sequence of matrices A_n if for each $\varepsilon > 0$ there exists a positive integer N such that if $n \geq N$, then $\| A - A_n \| < \varepsilon$.

THEOREM 4.1

Every convergent sequence of matrices A_n has a limit.

Proof.   The proof will follow from the fact that every convergent sequence of real numbers has a limit and the fact that $|a_{ij}^n - a_{ij}^m| \leq \| A_n - A_m \|$, where a_{ij}^p is the element in the ith row, jth column of the pth matrix in the sequence. Hence, if A_n converges, so does a_{ij}^n. Let $\lim_{n \to \infty} a_{ij}^n = a_{ij}, 1 \leq i \leq r, 1 \leq j \leq r$, and let $A = [a_{ij}]$. For $\varepsilon > 0$, choose N_{ij} such that $|a_{ij} - a_{ij}^n| < \varepsilon/r^2$ for $n \geq N_{ij}$, and let $N = \max_{ij} N_{ij}$. Then $\| A - A_n \| = \sum_{i,j} |a_{ij} - a_{ij}^n| < r^2 \varepsilon/r^2 \leq \varepsilon$ for $n \geq N$.

Given a sequence of matrices A_n, we can form another sequence (called the *sequence of partial sums*) by defining $S_n = A_1 + \cdots + A_n$. We denote the sequence $\{S_n\}$ by $\sum_{i=1}^{n} A_i$ and call $\sum_{i=1}^{\infty} A_i$ an infinite series. If $\lim_{n \to \infty} S_n = S$, then the series is said to *converge* and its sum is defined to be S. If $\{S_n\}$ does not converge, the series is said to *diverge*, and the sum is not defined. It is important to note that we can often show that a series converges without being able to find the limit. For example, we could investigate the (real) infinite series

$$\sum_{0}^{\infty} \frac{x^n}{n!},$$

or the sequence of partial sums

$$S_n = 1 + x + \frac{x^2}{2!} + \cdots + \frac{x^n}{n!}$$

and deduce that it converges. Then we could define a new function by

$$e^x = \sum_{0}^{\infty} \frac{x^n}{n!}.$$

All of the common properties of the exponential can be deduced from this series. While this approach is not commonly used in calculus for the real exponential, it is often used to define the exponential of a complex number. We will take this approach to define the exponential of a matrix and then deduce some of its properties through the limiting process. The series of interest is

$$I + A + \frac{A^2}{2!} + \cdots + \frac{A^n}{n!} + \cdots . \tag{4.1}$$

THEOREM 4.2

The series (4.1) is convergent.

Proof.   The nth partial sum is $S_n = \sum_{k=0}^{n} A^k/k!$, so

$$\| S_n - S_m \| = \left\| \sum_{k=0}^{n} \frac{A^k}{k!} - \sum_{k=0}^{m} \frac{A^k}{k!} \right\| = \left\| \sum_{k=m+1}^{n} \frac{A^k}{k!} \right\| ,$$

where we have chosen labeling so that $n > m$. By property (iii) of the norm, the last quantity above is

$$\leq \sum_{k=m+1}^{n} \frac{\| A^k \|}{k!} \leq \sum_{k=m+1}^{n} \frac{\| A \|^k}{k!} .$$

Since $\| A \|$ is a real number, the right-hand side is a part of the convergent series of real numbers

$$e^{\| A \|} = \sum_{k=0}^{\infty} \frac{\| A \|^k}{k!} . \tag{4.2}$$

Hence, since (4.2) is convergent, if $\varepsilon > 0$, there is an N such that for $m \geq N$,

$$\sum_{k=m+1}^{\infty} \frac{\| A \|^k}{k!} < \varepsilon .$$

This is sufficient to prove that $\{ S_n \}$ is convergent. It has a limit, by Theorem 4.1.

The sum of this series is denoted e^A, that is,

$$e^A = \sum_{k=0}^{\infty} \frac{A^k}{k!} . \tag{4.3}$$

Similarly, for a real number t,

$$e^{At} = \sum_{k=0}^{\infty} \frac{A^k t^k}{k!}.$$

Since each entry of e^{At} is defined as a convergent power series, it is differentiable (and hence continuous and integrable) and may be differentiated term by term. The pth term of the series is $A^p t^p / p!$. Hence, for $p \geq 1$ (for $p = 0$, $(d/dt)I = 0$, the derivative of a constant matrix is the null matrix) making use of (2.5),

$$\frac{d}{dt} \frac{A^p}{p!} t^p = \left[\frac{d}{dt} \frac{a_{ij}^p t^p}{p!} \right] = \left[p \frac{a_{ij}^p t^{p-1}}{p!} \right] = p \frac{A^p t^{p-1}}{p!}$$

$$= A \frac{A^{p-1} t^{p-1}}{(p-1)!},$$

where a_{ij}^p is the i-jth element of A^p. Hence,

$$(e^{At})' = \sum_{k=1}^{\infty} A \frac{A^{k-1} t^{k-1}}{(k-1)!} = A \sum_{k=0}^{\infty} \frac{A^k t^k}{k!} = A e^{At}.$$

THEOREM 4.3

$$(e^{At})' = A e^{At} = e^{At} A.$$

A matrix A is said to be similar to a matrix B if there exists a nonsingular matrix T such that $T^{-1} A T = B$. Similar matrices are related in ways that are useful to us in the study of differential equations. Here we note one such property involving the exponential of a matrix of the form $e^{TAT^{-1}}$. Since

$$(TAT^{-1})^n = \underbrace{(TAT^{-1})(TAT^{-1}) \cdots (TAT^{-1})}_{n \text{ times}},$$

$$(TAT^{-1})^n = TAT^{-1} TAT^{-1} \cdots TAT^{-1}$$

$$= TA^n T^{-1}.$$

If S_n is the nth partial sum of $e^{TAT^{-1}}$, that is,

$$S_n = \sum_{i=0}^{n} \frac{(TAT^{-1})^i}{i!},$$

then

$$S_n = \sum_{i=0}^{n} \frac{T A^i T^{-1}}{i!}$$

$$= T \left(\sum_{i=0}^{n} \frac{A^i}{i!} \right) T^{-1}.$$

From this it follows that

$$\lim_{n \to \infty} S_n = T e^A T^{-1};$$

that is,

$$e^{T A T^{-1}} = T e^A T^{-1}. \tag{4.4}$$

We need an additional fact about e^A, whose proof we defer to the exercises.

THEOREM 4.4

det $e^M \neq 0$ for any matrix M; that is, e^M is always nonsingular.

EXERCISES

1. Let $x = \begin{pmatrix} a_1 \\ a_2 \end{pmatrix}$ be a vector in R^2. Describe (geometrically) the set of points x such that $\|x\| = 1$ and such that $\|x\| < 1$.

2. If A is an $n \times n$ matrix and x is an n-vector, show that $\|Ax\| \leq \|A\| \|x\|$. (*Hint:* Use the definition of multiplication.)

3. Compute e^A by summing (4.3) when $A =$

   (a) $\begin{bmatrix} 1 & 0 \\ 0 & 2 \end{bmatrix}$ (b) $\begin{bmatrix} 1 & 0 & 0 \\ 0 & 1 & 0 \\ 0 & 0 & 3 \end{bmatrix}$ (c) $\begin{bmatrix} 2 & 0 & 0 \\ 0 & 2 & 0 \\ 0 & 0 & 2 \end{bmatrix}$

4. Show that if $A = \begin{bmatrix} \lambda_1 & 0 & 0 \\ 0 & \lambda_2 & 0 \\ 0 & 0 & \lambda_3 \end{bmatrix}, e^A = \begin{bmatrix} e^{\lambda_1} & 0 & 0 \\ 0 & e^{\lambda_2} & 0 \\ 0 & 0 & e^{\lambda_3} \end{bmatrix}.$

5. Let $A = \begin{bmatrix} 0 & 1 & 0 \\ 0 & 0 & 1 \\ 0 & 0 & 0 \end{bmatrix}$. Compute A^2, A^3, and A^4.

6. Find e^A where A is as in Exercise 5.

7. If A and B are $n \times n$ matrices such that $AB = BA$, show that $e^A e^B = e^{A+B}$.

8. Combine Exercises 4 and 7 and the definition (4.3) to find e^A where $A =$

(a) $\begin{bmatrix} 1 & 1 \\ 0 & 1 \end{bmatrix}$ (b) $\begin{bmatrix} 2 & 1 & 0 \\ 0 & 2 & 0 \\ 0 & 0 & 1 \end{bmatrix}$ (c) $\begin{bmatrix} 1 & 1 & 0 \\ 0 & 1 & 0 \\ 0 & 0 & 1 \end{bmatrix}$

9. Show that $(e^A)^{-1} = e^{-A}$. (*Hint:* See Exercise 7.)

10. Show that $\det e^M \neq 0$, for any M. (*Hint:* $e^M e^{-M} = I$, by Exercise 9.)

11. Let $A(t)$ be a continuous matrix defined on an interval I. Show that the scalar function $\|A(t)\|$ is a continuous function.

12. Let $v = \begin{pmatrix} v_1 \\ \vdots \\ v_n \end{pmatrix}$ be a vector in R^n. Let $N(v) = \max_i [|v_1|, |v_2|, \ldots, |v_n|]$. Show that $N(v)$ satisfies properties (i), (ii), and (iii) of the listed properties for the norm of a matrix.

13. Let $N(v)$ be as in Exercise 12. Show that

$$N(v) \leq \|v\| \leq nN(v).$$

14. Let $A(t)$ be a continuous matrix defined on an interval $[a, b]$. Show that

$$\left\| \int_a^b A(t)\, dt \right\| \leq \int_a^b \|A(t)\|\, dt.$$

(*Hint:* Use the definition of $\|\quad\|$ and a similar property of absolute value for scalar functions.)

5. The Constant Coefficient Case: Real and Distinct Eigenvalues

The theory developed in Section 3 exhibited the rich structure of linear systems of differential equations. Unfortunately, that theory gave no clues as to how we construct the fundamental matrix on which that theory depends. This is not unexpected since, even for the simple scalar equation $y'' + a(t)y = 0$, there is no "formula" for the general solution. The class of equations that we can actually solve is far smaller than that to which the theory of Section 3 applies. However, using the material developed in Section 4, it is possible to construct solutions to those systems where the coefficient matrix is constant, that is, systems of the form

$$y' = Ay$$

where A is a constant matrix. The amount of work required to do this is somewhat more than in the case of scalar equations that the reader may have encountered previously, but the added difficulties are those of linear algebra rather than of differential equations. Several new concepts from linear algebra appear here, particularly the important concepts of eigenvalue and eigenvector. The first theorem shows the importance of the concept of the exponential of a matrix, developed in Section 4. The remainder of this section is devoted to making this idea constructive.

THEOREM 5.1

   Let A be a constant matrix. A fundamental matrix Φ for

$$y' = Ay \tag{5.1}$$

   is given by

$$\Phi = e^{At}. \tag{5.2}$$

Proof. From Theorem 4.3, $(e^{At})' = Ae^{At}$, so $\Phi'(t) = A\Phi(t)$. Furthermore, $\det e^{At} \neq 0$ (Theorem 4.4), and thus Φ is a fundamental matrix.

The matrix e^{At} is, however, not readily accessible, for since

$$e^{At} = I + At + \frac{A^2}{2!}t^2 + \cdots \tag{5.3}$$

it is necessary to sum the series. There is one tractable case (see Exercise 3 of Section 4), the case where the matrix is diagonal, that is, $a_{ii} = \lambda_i$, $a_{ij} = 0$, $i \neq j$, or

$$A = \begin{bmatrix} \lambda_1 & 0 & \cdots & & 0 \\ 0 & \lambda_2 & \cdots & & 0 \\ 0 & 0 & \lambda_3 & \cdots & 0 \\ 0 & 0 & 0 & & \lambda_n \end{bmatrix}, \tag{5.4}$$

for then A^2 is

$$\begin{bmatrix} \lambda_1^2 & 0 & \cdots & 0 \\ 0 & \lambda_2^2 & \cdots & 0 \\ 0 & \cdots & & \lambda_n^2 \end{bmatrix},$$

A^3 has λ_i^3 on the diagonal, and so on. The series (4.3) may be summed to obtain

$$e^{At} = \begin{bmatrix} e^{\lambda_1 t} & 0 & \cdots & & 0 \\ 0 & e^{\lambda_2 t} & 0 & \cdots & 0 \\ \vdots & & & & \\ 0 & & \cdots & & e^{\lambda_n t} \end{bmatrix}.$$

This is not surprising, of course, since the system (4.1) in this case is

$$x_1' = \lambda_1 x_1$$
$$x_2' = \lambda_2 x_2$$
$$\vdots$$
$$x_n' = \lambda_n x_n.$$

The system is *uncoupled* (each equation does not involve any other) and each equation may be solved directly to yield $x_i(t) = e^{\lambda_i t}$.

If A is not diagonal, it still may be the case that there is a matrix T such that TAT^{-1} has the form (5.4). (Recall that this means that A is similar to a diagonal matrix.) From (4.4) it follows that

$$Te^{At}T^{-1} = e^{TAT^{-1}}t. \tag{5.5}$$

Thus, if we can find a matrix T and the λ's resulting after the transformation, we can, of course, recover e^{At}. Saying this another way, let $B = TAT^{-1}$ and have the form (5.4). Then the solution of

$$y' = By$$

can be found, since e^{Bt} can be computed. From $e^{At} = T^{-1}e^{Bt}T$, we have e^{At}. Rather than attempt to find the T, we reason as follows: e^{Bt} has entries $e^{\lambda_i t}$. Premultiplication (multiplication on the left) by T^{-1} and postmultiplication (multiplication on the right) by T rearranges and combines these. Suppose we disregard B and simply look for a solution of (5.1) of the form

$$y = \begin{bmatrix} c_1 e^{\lambda_i t} \\ c_2 e^{\lambda_i t} \\ \vdots \\ c_n e^{\lambda_i t} \end{bmatrix} = e^{\lambda_i t} \begin{bmatrix} c_1 \\ \vdots \\ c_n \end{bmatrix} = e^{\lambda_i t} c,$$

where λ_i is one of the diagonal elements of $B = TAT^{-1}$. Then

$$y' = \begin{bmatrix} \lambda_i c_i e^{\lambda_i t} \\ \vdots \\ \lambda_i c_n e^{\lambda_i t} \end{bmatrix} = \lambda_i y,$$

or

$$\lambda_i y = Ay,$$

which we write as

$$(A - \lambda_i I)y = 0. \tag{5.6}$$

If y is not to be identically zero, then $A - \lambda_i I$ must be singular (Theorem 2.5). The (complex) numbers λ such that

$$\det(A - \lambda I) = 0,$$

are called the *eigenvalues* of the matrix A. Vectors c, not identically zero, such that

$$(A - \lambda I)c = 0,$$

are called *eigenvectors*. Equation (5.6) says that λ_i must be an eigenvalue of A and substitution for y gives

$$(A - \lambda_i I)e^{\lambda_i t}c = 0,$$

or

$$(A - \lambda_i I)c = 0, \tag{5.7}$$

since $e^{\lambda_i t} \neq 0$. Equation (5.7) says that c must be an eigenvector of A. Since the λ_i are fixed, that is, they are the diagonal elements of B, it would seem that we have no hope of satisfying (5.6). (Equation (5.7) can be satisfied, since the constant vector c has been arbitrary up to now.) This matter is resolved in the following.

THEOREM 5.2
   If $TAT^{-1} = B$, then A and B have the same eigenvalues.

Proof. $B - \lambda I = TAT^{-1} - \lambda TT^{-1}$

$$= T(A - \lambda I)T^{-1}.$$

Thus, using Theorem 2.3,

$$\det(B - \lambda I) = (\det T)(\det(A - \lambda I))(\det T^{-1}).$$

Now both $\det T$ and $\det T^{-1} \neq 0$, since T^{-1} exists. Thus, $\det(B - \lambda I) = 0$ if and only if $\det(A - \lambda I) = 0$ or A and B have the same eigenvalues.

Note that for a diagonal matrix B, the eigenvalues are the n entries. Clearly, $\det(B - \lambda I)$ is a polynomial of degree n in λ; hence, so is $\det(A - \lambda I)$. This polynomial is called the *characteristic polynomial*.

Thus, if λ is an eigenvalue of A and if c is chosen as an eigenvector, $e^{\lambda_i t}c$ is a solution. The following theorem summarizes this argument.

THEOREM 5.3

If A is a constant matrix, λ an eigenvalue of A, and c a corresponding eigenvector, then $y = e^{\lambda t}c$ is a solution of (5.1)

Since A has n eigenvalues, we can find n such solutions, and it would seem then that we have found the columns for a fundamental matrix. The difficulty, however, is that the eigenvalues are not necessarily distinct and the eigenvectors corresponding to a repeated eigenvalue may not be linearly independent. (Eigenvectors corresponding to distinct eigenvalues are always linearly independent.) If this occurs, we have not found n linearly independent column vectors to make a fundamental matrix. The analysis in this case is a good bit more complicated, and we defer it for the moment. However, it is the case that if all of the eigenvalues of A are distinct, then A is similar to a diagonal matrix, so the n solutions obtained actually are linearly independent, and a fundamental matrix has been found.

THEOREM 5.4

Let A be a constant $n \times n$ matrix with distinct eigenvalues $\lambda_1, \ldots, \lambda_n$ and let $c_1, \ldots, c_n$ be corresponding eigenvectors. Then a fundamental matrix for (5.1) is given by

$$\Phi(t) = [e^{\lambda_1 t}c_1, e^{\lambda_2 t}c_2, \ldots, e^{\lambda_n t}c_n].$$

We illustrate the foregoing analysis with some examples, where the λ_i's are real numbers. First, consider the system

$$x_1' = 2x_1 - x_2$$
$$x_2' = 4x_2$$
$$x_3' = 2x_1 + 5x_2 + 3x_3$$

or, in matrix form,

$$\begin{bmatrix} x_1 \\ x_2 \\ x_3 \end{bmatrix}' = \begin{bmatrix} 2 & -1 & 0 \\ 0 & 4 & 0 \\ 2 & 5 & 3 \end{bmatrix} \begin{bmatrix} x_1 \\ x_2 \\ x_3 \end{bmatrix}. \tag{5.8}$$

The eigenvalues are the solutions of

$$\det\left(\begin{bmatrix} 2 & -1 & 0 \\ 0 & 4 & 0 \\ 2 & 5 & 3 \end{bmatrix} - \lambda \begin{bmatrix} 1 & 0 & 0 \\ 0 & 1 & 0 \\ 0 & 0 & 1 \end{bmatrix} \right) = 0$$

or

$$\det \begin{bmatrix} 2-\lambda & -1 & 0 \\ 0 & 4-\lambda & 0 \\ 2 & 5 & 3-\lambda \end{bmatrix} = 0.$$

Expansion gives

$$(2-\lambda)\begin{bmatrix} 4-\lambda & 0 \\ 5 & 3-\lambda \end{bmatrix} + 2\begin{bmatrix} -1 & 0 \\ 4-\lambda & 0 \end{bmatrix} = 0$$

or

$$(2-\lambda)(4-\lambda)(3-\lambda) = 0.$$

Thus, the eigenvalues are $\lambda_1 = 2, \lambda_2 = 3$, and $\lambda_3 = 4$, all distinct. Hence, if we can find the eigenvectors, we can find a fundamental matrix.

An eigenvector can be determined by solving

$$\begin{bmatrix} 2-\lambda & -1 & 0 \\ 0 & 4-\lambda & 0 \\ 2 & 5 & 3-\lambda \end{bmatrix} \begin{bmatrix} c_1 \\ c_2 \\ c_3 \end{bmatrix} = 0$$

for $\lambda = 2, 3, 4$. If $\lambda = 2$, this becomes

$$\begin{bmatrix} 0 & -1 & 0 \\ 0 & 2 & 0 \\ 2 & 5 & 1 \end{bmatrix}\begin{bmatrix} c_1 \\ c_2 \\ c_3 \end{bmatrix} = 0.$$

This yields $c_2 = 0$ and

$$2c_1 + c_3 = 0.$$

Thus, an eigenvector (clearly there are infinitely many, since a constant multiple of an eigenvector satisfies the defining equation) is

$$c = \begin{bmatrix} 1 \\ 0 \\ -2 \end{bmatrix}.$$

For $\lambda = 3$, the system becomes

$$\begin{bmatrix} -1 & -1 & 0 \\ 0 & 1 & 0 \\ 2 & 5 & 0 \end{bmatrix}\begin{bmatrix} c_1 \\ c_2 \\ c_3 \end{bmatrix} = 0$$

or

$$c_1 + c_2 = 0$$
$$c_2 = 0$$
$$2c_1 + 5c_2 = 0.$$

Here c_1 and c_2 are zero, and since c_3 does not appear in the equations, it may be chosen arbitrarily. Hence, an eigenvector is

$$\begin{bmatrix} 0 \\ 0 \\ 1 \end{bmatrix}.$$

Finally, for $\lambda = 4$, the equations are

$$2c_1 + c_2 = 0$$
$$2c_1 + 5c_2 - c_3 = 0.$$

Setting $c_2 = 2$ (arbitrarily), $c_1 = -1$ and $c_3 = 2c_1 + 5c_2 = -2 + 10 = 8$. Thus, a final eigenvector is

$$\begin{bmatrix} -1 \\ 2 \\ 8 \end{bmatrix}.$$

Three linearly independent solutions are

$$\begin{bmatrix} 1 \\ 0 \\ -2 \end{bmatrix} e^{2t}, \quad \begin{bmatrix} 0 \\ 0 \\ 1 \end{bmatrix} e^{3t}, \quad \begin{bmatrix} -1 \\ 2 \\ 8 \end{bmatrix} e^{4t}$$

and a fundamental matrix corresponding to (5.8) is

$$\Phi(t) = \begin{bmatrix} e^{2t} & 0 & -e^{4t} \\ 0 & 0 & 2e^{4t} \\ -2e^{2t} & e^{3t} & 8e^{4t} \end{bmatrix}. \tag{5.9}$$

We now have $\det \Phi(t) = -e^{3t}(e^{2t})(2e^{4t}) = -2e^{9t}$, which, as it must be, is $\neq 0$.

As another example, consider the system

$$x_1' = x_1 + x_3$$
$$x_2' = x_1 + 2x_2 + 3x_3$$
$$x_3' = 3x_3$$

which, in matrix form, is

$$x' = \begin{pmatrix} 1 & 0 & 1 \\ 1 & 2 & 3 \\ 0 & 0 & 3 \end{pmatrix} x,$$

with $x = \begin{pmatrix} x_1 \\ x_2 \\ x_3 \end{pmatrix}$. The eigenvalues are the solutions of

$$\det(A - \lambda I) = \det \begin{pmatrix} 1 - \lambda & 0 & 1 \\ 1 & 2 - \lambda & 3 \\ 0 & 0 & 3 - \lambda \end{pmatrix}$$

$$= (1 - \lambda)(2 - \lambda)(3 - \lambda) = 0$$

or $\lambda_1 = 1$, $\lambda_2 = 2$, $\lambda_3 = 3$. For $\lambda = \lambda_1 = 1$ an eigenvector is the solution of

$$(A - I)c = \begin{pmatrix} 0 & 0 & 1 \\ 1 & 1 & 3 \\ 0 & 0 & 2 \end{pmatrix} \begin{pmatrix} c_1 \\ c_2 \\ c_3 \end{pmatrix} = 0.$$

In equation form, this is

$$c_3 = 0$$
$$c_1 + c_2 + 3c_3 = 0$$
$$2c_3 = 0.$$

A solution of this system of linear equations is given by $c_1 = 1$, $c_2 = -1$, $c_3 = 0$ and one solution of the system of differential equations takes the form

$$x_1(t) = \begin{pmatrix} 1 \\ -1 \\ 0 \end{pmatrix} e^t.$$

For $\lambda = \lambda_2 = 2$, an eigenvector is the solution of

$$(A - 2I)c = \begin{pmatrix} -1 & 0 & 1 \\ 1 & 0 & 3 \\ 0 & 0 & 1 \end{pmatrix} \begin{pmatrix} c_1 \\ c_2 \\ c_3 \end{pmatrix} = 0$$

or

$$-c_1 + c_3 = 0$$
$$c_1 + c_3 = 0$$
$$c_3 = 0.$$

One solution of these equations is $c_1 = c_3 = 0$, $c_2 = 1$, yielding a solution to the differential equation

$$x_2(t) = e^{2t} \begin{pmatrix} 0 \\ 1 \\ 0 \end{pmatrix}.$$

Finally, for $\lambda = \lambda_3 = 3$, an eigenvector is a solution of

$$(A - 3I)c = \begin{pmatrix} -2 & 0 & 1 \\ 1 & -1 & 3 \\ 0 & 0 & 0 \end{pmatrix} \begin{pmatrix} c_1 \\ c_2 \\ c_3 \end{pmatrix} = 0$$

or

$$-2c_1 + c_3 = 0$$
$$c_1 - c_2 + 3c_3 = 0.$$

Since $c_3 = 2c_1$, then

$$7c_1 = c_2.$$

Choosing (arbitrarily) $c_1 = 1$ yields $c_2 = 7$ and $c_3 = 2$. Hence, a third solution of the system of differential equations is given by

$$x_3 = e^{3t} \begin{pmatrix} 1 \\ 7 \\ 2 \end{pmatrix}.$$

Using these three (linearly independent) solutions, a fundamental matrix Φ takes the form

$$\Phi(t) = \begin{pmatrix} e^t & 0 & e^{3t} \\ -e^t & e^{2t} & 7e^{3t} \\ 0 & 0 & 2e^{3t} \end{pmatrix}.$$

Suppose that, in addition to solving the system of differential equations, we want the solution through the vector $\begin{pmatrix} 1 \\ 1 \\ 0 \end{pmatrix}$ at time $t = 0$. Since $\Phi(t)$ is a fundamental matrix, any solution takes the form $\Phi(t)c$, for some vector c. To fit the initial condition it is then necessary to solve

$$\Phi(0)c = \begin{pmatrix} 1 \\ 1 \\ 0 \end{pmatrix}$$

for $c = \begin{pmatrix} c_1 \\ c_2 \\ c_3 \end{pmatrix}$. Thus, we must solve

$$\begin{pmatrix} 1 & 0 & 1 \\ -1 & 1 & 7 \\ 0 & 0 & 2 \end{pmatrix} \begin{pmatrix} c_1 \\ c_2 \\ c_3 \end{pmatrix} = \begin{pmatrix} 1 \\ 1 \\ 0 \end{pmatrix}.$$

In equation form this is,

$$c_1 + c_3 = 1$$
$$-c_1 + c_2 + 7c_3 = 1$$
$$2c_3 = 0$$

or

$$c_3 = 0$$
$$c_1 = 1$$
$$c_2 = 2.$$

The desired solution, $y(t)$, is then given by

$$y(t) = \begin{pmatrix} e^t & 0 & e^{3t} \\ -e^t & e^{2t} & 7e^{3t} \\ 0 & 0 & 2e^{3t} \end{pmatrix} \begin{pmatrix} 1 \\ 2 \\ 0 \end{pmatrix}$$

$$= \begin{pmatrix} e^t \\ 2e^{2t} - e^t \\ 0 \end{pmatrix}.$$

Finally, consider the system

$$\begin{bmatrix} x_1 \\ x_2 \\ x_3 \end{bmatrix}' = \begin{bmatrix} 1 & 3 & -2 \\ 0 & 1 & 4 \\ 0 & 0 & 1 \end{bmatrix} \begin{bmatrix} x_1 \\ x_2 \\ x_3 \end{bmatrix}.$$

The eigenvalues are the roots of

$$\det \begin{bmatrix} 1 - \lambda & 3 & -2 \\ 0 & 1 - \lambda & 4 \\ 0 & 0 & 1 - \lambda \end{bmatrix} = 0$$

or

$$(1 - \lambda)(1 - \lambda)(1 - \lambda) = 0;$$

that is, $\lambda = 1$ is a triple root. An eigenvector can be found by solving

$$
\begin{bmatrix} 0 & 3 & -2 \\ 0 & 0 & 4 \\ 0 & 0 & 0 \end{bmatrix} \begin{bmatrix} c_1 \\ c_2 \\ c_3 \end{bmatrix} = 0.
$$

An eigenvector is

$$
c = \begin{bmatrix} 1 \\ 0 \\ 0 \end{bmatrix}
$$

and one solution is

$$
x = \begin{bmatrix} e^t \\ 0 \\ 0 \end{bmatrix}.
$$

To find two additional linearly independent solutions requires additional analysis, which is presented in Section 7. It should be noted here that a theorem from linear algebra states that if the matrix A is similar to a diagonal matrix, there will always be enough (that is, n) linearly independent eigenvectors to successfully carry out the procedure for finding a fundamental matrix. Two particular cases are worthy of note. For each eigenvalue there is always one nontrivial eigenvector, so the procedure above will work (as noted above) if there are n distinct eigenvalues. If the matrix A satisfies the additional property that $a_{ij} = a_{ji}$—such matrices are said to be *symmetric*—then A is similar to a diagonal matrix and there will be "enough" eigenvectors to find a fundamental matrix.

EXERCISES

1. Compute eigenvalues of the following matrices:

(a) $\begin{pmatrix} 1 & 2 \\ 0 & 2 \end{pmatrix}$

(b) $\begin{pmatrix} 1 & 2 \\ 3 & 4 \end{pmatrix}$

(c) $\begin{pmatrix} -1 & -1 \\ 2 & 4 \end{pmatrix}$

(d) $\begin{pmatrix} 2 & 4 \\ 1 & 1 \end{pmatrix}$

(e) $\begin{pmatrix} 3 & -1 & 0 \\ 2 & 0 & 0 \\ 2 & -2 & 3 \end{pmatrix}$

(f) $\begin{pmatrix} 1 & 0 & 0 \\ 0 & 1 & 0 \\ 0 & 0 & 1 \end{pmatrix}$

2. Find eigenvectors corresponding to the eigenvalues found in Exercise 1.

3. Find a fundamental matrix for $x' = Ax$ for each A given in Exercise 1.

4. Find the solution of $x' = Ax$, $x(0) = \begin{pmatrix} 1 \\ 1 \end{pmatrix}$, where A is given in Exercise 1(a)–(d).

5. Find the solution of $x' = Ax$, $x(0) = \begin{pmatrix} 1 \\ 0 \\ 1 \end{pmatrix}$, where A is given in Exercise 1(f).

6. Show that the matrices $\begin{bmatrix} 1 & 2 \\ 3 & 4 \end{bmatrix}$ and $\begin{bmatrix} -3 & 11 \\ -2 & 8 \end{bmatrix}$ are similar. (*Hint:* Let $T = [t_{ij}]$ and try to solve $AT = TB$.)

7. Let $A = \begin{pmatrix} 1 & 2 \\ 3 & 4 \end{pmatrix}$. Let t_1 and t_2 be the eigenvectors found in Exercise 2(b). Define $T = [t_1, t_2]$, a 2×2 matrix. Compute $T^{-1}AT$ and TAT^{-1}.

8. Let $A = \begin{pmatrix} 2 & 4 \\ 1 & 4 \end{pmatrix}$. Find two linearly independent eigenvectors and repeat Exercise 7.

9. For A given in equation (5.8), (5.9) is not e^{At}. (Why?) Find a matrix C such that $\Phi(t)C = e^{At}$. (*Hint:* It is sufficient that $\Phi(0)C = I$.)

10. Show that eigenvectors corresponding to distinct eigenvalues of a matrix A are linearly independent. (*Hint:* Suppose eigenvectors v_1 and v_2 correspond to eigenvalues $\lambda_1, \lambda_2, \lambda_1 \neq \lambda_2$, and that $c_1 v_1 + c_2 v_2 = 0$. Apply A to both sides.)

6. The Constant Coefficient Case: Complex and Distinct Eigenvalues

Nowhere in the development of the theory in Section 5 was any explicit use made of the assumption that the eigenvalues of the matrix A were real numbers. If some of the eigenvalues λ_i turn out to be complex numbers, then the corresponding eigenvectors, c_i, will contain complex entries, but $e^{\lambda_i t} c_i$ will still be a solution. For most problems with real coefficients, we are interested in having real-valued solutions. Since all initial conditions can be satisfied, given a fundamental matrix Φ, real solutions are of the form $\Phi(t)c$, where $\Phi(t)$ and c may have complex entries. Representing a real vector as the product of a matrix with complex entries and a constant vector with complex entries is, at least, inelegant and frequently may be awkward. For this reason we seek a way to find a real fundamental matrix. That this can always be done is a consequence of the following theorem.

THEOREM 6.1

If $\varphi(t)$ is a solution of

$$x' = Ax, \tag{6.1}$$

where A is a constant matrix with real-valued entries, then the real part of $\varphi(t)$ (written Re $\varphi(t)$) and the imaginary part of $\varphi(t)$ (written Im $\varphi(t)$) are both solutions of (6.1).

Proof. The complex-valued function $\varphi(t)$ can be written as

$$\varphi(t) = u(t) + iv(t)$$

where $u(t)$ and $v(t)$ are real-valued functions ($u(t) = \operatorname{Re}\varphi(t)$, $v(t) = \operatorname{Im}\varphi(t)$). Since $\varphi(t)$ is a solution of (6.1),

$$(u(t) + iv(t))' = A(u(t) + iv(t)),$$

or, using the distributive law for matrices, and the fact that differentiation is linear,

$$u'(t) + iv'(t) = Au(t) + iAv(t).$$

$Au(t)$, $Av(t)$, $u'(t)$, and $v'(t)$ are real-valued vectors, and two complex vectors can be equal if and only if the real parts and the imaginary parts are equal. Hence, it must be the case that

$$u'(t) = Au(t), \qquad \text{all } t,$$

and

$$v'(t) = Av(t), \qquad \text{all } t.$$

Thus, $u(t)$ and $v(t)$ solve (6.1).

Returning to our original discussion, if λ_i is a complex eigenvalue of A and c_i is the corresponding complex eigenvector, then $\operatorname{Re}(e^{\lambda_i t}c_i)$ and $\operatorname{Im}(e^{\lambda_i t}c_i)$ are solutions. Making use of Euler's formula,

$$e^{i\theta} = \cos\theta + i\sin\theta, \tag{6.2}$$

we can be more explicit; let $\lambda_i = \alpha + i\beta$ and $c = a + ib$, α, β real numbers, a, b real vectors. Then

$$\operatorname{Re}[e^{(\alpha+i\beta)t}(a + ib)] = e^{\alpha t}[(\cos(\beta t))a - (\sin(\beta t))b] \tag{6.3}$$

and

$$\text{Im}\,[e^{(\alpha+i\beta)t}(a+ib)] = e^{\alpha t}[(\sin(\beta t))a + (\cos(\beta t))b]. \tag{6.4}$$

It is not difficult to show that these two vectors are linearly independent.

At first glance it would seem that from one solution, two linearly independent vectors have been created. This is not the case, and we explore this point in somewhat more detail. First of all, since the matrix A has real coefficients, $\det(A - \lambda I) = p(\lambda)$ is a polynomial with real coefficients. Let

$$p(\lambda) = \lambda^n + a_1 \lambda^{n-1} + \cdots + a_n.$$

Let λ be any complex number. Then (denoting the complex conjugate of a complex number z by $\bar{z}$),

$$\overline{p(\lambda)} = \overline{\lambda^n + a_1 \lambda^{n-1} + \cdots + a_n}$$
$$= \overline{\lambda^n} + \overline{a_1 \lambda^{n-1}} + \cdots + \overline{a_n}$$
$$= \bar{\lambda}^n + a_1 \bar{\lambda}^{n-1} + \cdots + a_n,$$

since $\bar{a}_i = a_i$ and $\overline{\lambda^n} = \bar{\lambda}^n$. Thus, $\overline{p(\lambda)} = p(\bar{\lambda})$. If λ_i is a complex eigenvalue, then $p(\lambda_i) = 0$, which implies that $p(\bar{\lambda}_i) = 0$ or that $\bar{\lambda}_i$ is an eigenvalue. Thus, complex eigenvalues occur as complex-conjugate pairs.

Now let c_i be an eigenvector corresponding to the complex eigenvalue λ_i. Then, since $(A - \lambda_i I)c_i = 0$,

$$\overline{(A - \lambda_i I)c_i} = 0$$

or

$$(A - \bar{\lambda}_i I)\bar{c}_i = 0.$$

Since $\bar{\lambda}_i$ is an eigenvalue of A, $\bar{c}_i$ is an eigenvector. Thus, knowing a complex eigenvalue λ and its corresponding eigenvector c lets us determine a second eigenvalue-eigenvector pair, $\bar{\lambda}, \bar{c}$. In effect, taking the real and imaginary parts of a complex solution $e^{\lambda_i t}c_i$ amounts to using both λ_i and $\bar{\lambda}_i$ and c_i and $\bar{c}_i$ to find two real (linearly independent) solutions.

We illustrate the procedure with some examples. Consider the system

$$x' = \begin{pmatrix} 0 & -1 \\ 1 & 0 \end{pmatrix} x.$$

$\det(A - \lambda I) = \det\begin{pmatrix} -\lambda & -1 \\ 1 & -\lambda \end{pmatrix} = 0$ yields $\lambda^2 + 1 = 0$ or $\lambda = \pm i$. Fix $\lambda = i$.
To find an eigenvector, it is necessary to solve

$$\begin{pmatrix} -i & -1 \\ 1 & -i \end{pmatrix}\begin{pmatrix} c_1 \\ c_2 \end{pmatrix} = 0$$

or

$$-ic_1 - c_2 = 0$$

$$c_1 - ic_2 = 0.$$

Then $c_1 = 1$, $c_2 = -i$ is a nontrivial solution to this linear system of equations, so a solution vector is given by

$$\varphi(t) = e^{it}\begin{pmatrix} 1 \\ -i \end{pmatrix}.$$

Making use of Euler's formula, $e^{i\theta} = \cos(\theta) + i\sin(\theta)$,

$$\varphi(t) = (\cos(t) + i\sin(t))\begin{pmatrix} 1 \\ -i \end{pmatrix}$$

$$= \begin{pmatrix} \cos(t) \\ \sin(t) \end{pmatrix} + i\begin{pmatrix} \sin(t) \\ -\cos(t) \end{pmatrix}.$$

Thus, two solutions are given by

$$\text{Re}\,\varphi(t) = \begin{pmatrix} \cos(t) \\ \sin(t) \end{pmatrix},$$

$$\text{Im}\,\varphi(t) = \begin{pmatrix} \sin(t) \\ -\cos(t) \end{pmatrix}$$

and

$$\Phi(t) = \begin{pmatrix} \cos(t) & \sin(t) \\ \sin(t) & -\cos(t) \end{pmatrix}$$

is a fundamental matrix with real entries.

Consider now the system

$$x' = \begin{pmatrix} 1 & 1 & 0 \\ -1 & 1 & 0 \\ 1 & 0 & 1 \end{pmatrix} x.$$

Then $\det(A - \lambda I) = (1 - \lambda)((1 - \lambda)^2 + 1)$, or the eigenvalues are $\lambda = 1$ and

$\lambda = 1 \pm i$. Fixing $\lambda = 1$, we seek a nontrivial solution of

$$\begin{pmatrix} 0 & 1 & 0 \\ 1 & 0 & 0 \\ 1 & 0 & 0 \end{pmatrix} \begin{pmatrix} c_1 \\ c_2 \\ c_3 \end{pmatrix} = 0.$$

Thus, $c_2 = 0, c_1 = 0$, and c_3 is arbitrary, or $\begin{pmatrix} 0 \\ 0 \\ 1 \end{pmatrix}$ is an eigenvector and $e^t \begin{pmatrix} 0 \\ 0 \\ 1 \end{pmatrix}$ is

a solution. Fixing $\lambda = 1 + i$, it is necessary to solve

$$\begin{pmatrix} -i & 1 & 0 \\ -1 & -i & 0 \\ 1 & 0 & -i \end{pmatrix} \begin{pmatrix} c_1 \\ c_2 \\ c_3 \end{pmatrix} = 0.$$

This requires

$$-c_1 i + c_2 = 0$$
$$-c_1 - c_2 i = 0$$
$$c_1 - c_3 i = 0.$$

Setting $c_1 = 1$ (arbitrarily) produces $c_2 = i$ and $c_3 = -i$; or, we can conclude that

$$\varphi(t) = e^t(\cos(t) + i\sin(t)) \begin{pmatrix} 1 \\ i \\ -i \end{pmatrix}$$

is a solution. To find two real solutions it is only necessary to decompose this solution into its real and imaginary parts. A straightforward computation shows that

$$\varphi(t) = e^t \left[\begin{pmatrix} \cos(t) \\ -\sin(t) \\ \sin(t) \end{pmatrix} + i \begin{pmatrix} \sin(t) \\ \cos(t) \\ -\cos(t) \end{pmatrix} \right].$$

The two desired real solutions are

$$e^t \begin{pmatrix} \cos(t) \\ -\sin(t) \\ \sin(t) \end{pmatrix} \text{ and } e^t \begin{pmatrix} \sin(t) \\ \cos(t) \\ -\cos(t) \end{pmatrix},$$

and

$$\Phi = \begin{pmatrix} 0 & e^t \cos(t) & e^t \sin(t) \\ 0 & -e^t \sin(t) & e^t \cos(t) \\ e^t & e^t \sin(t) & -e^t \cos(t) \end{pmatrix}$$

is a real fundamental matrix.

We conclude with one additional example with all complex eigenvalues. Consider the system

$$x' = \begin{bmatrix} 1 & 1 & 0 & 1 \\ -1 & 1 & 0 & 1 \\ 0 & 0 & 2 & 1 \\ 0 & 0 & -1 & 2 \end{bmatrix} x.$$

The eigenvalues are the roots of

$$\det \begin{bmatrix} 1-\lambda & 1 & 0 & 1 \\ -1 & 1-\lambda & 0 & 1 \\ 0 & 0 & 2-\lambda & 1 \\ 0 & 0 & -1 & 2-\lambda \end{bmatrix} = 0.$$

Expanding this determinant yields the characteristic polynomial in the form

$$p(\lambda) = [(1-\lambda)^2 + 1][(2-\lambda)^2 + 1] = 0.$$

Thus, the eigenvalues are $1 \pm i$ and $2 \pm i$. To find an eigenvector corresponding to $1 - i$, we must solve the system

$$\begin{bmatrix} i & 1 & 0 & 1 \\ -1 & i & 0 & 1 \\ 0 & 0 & 1+i & 1 \\ 0 & 0 & -1 & 1+i \end{bmatrix} \begin{bmatrix} c_1 \\ c_2 \\ c_3 \\ c_4 \end{bmatrix} = 0.$$

This is the same as

$$ic_1 + c_2 + c_4 = 0$$
$$-c_1 + ic_2 + c_4 = 0$$
$$(1+i)c_3 + c_4 = 0$$
$$-c_3 + (1+i)c_4 = 0.$$

A solution of this linear algebraic system is given by the vector

$$\begin{bmatrix} i \\ 1 \\ 0 \\ 0 \end{bmatrix},$$

and a solution of the differential equation is

$$\begin{bmatrix} i \\ 1 \\ 0 \\ 0 \end{bmatrix} e^t[\cos(t) - i\sin(t)].$$

Taking the real and imaginary parts, we find two real solutions of the system

$$\begin{bmatrix} \sin(t) \\ \cos(t) \\ 0 \\ 0 \end{bmatrix} e^t, \qquad \begin{bmatrix} \cos(t) \\ -\sin(t) \\ 0 \\ 0 \end{bmatrix} e^t.$$

Now let $\lambda = 2 - i$ and seek an eigenvector by solving the system

$$\begin{bmatrix} -1 + i & 1 & 0 & 1 \\ -1 & -1 + i & 0 & 1 \\ 0 & 0 & i & 1 \\ 0 & 0 & -1 & i \end{bmatrix} \begin{bmatrix} c_1 \\ c_2 \\ c_3 \\ c_4 \end{bmatrix} = 0.$$

This is the same as

$$(-1 + i)c_1 + c_2 + c_4 = 0$$
$$-c_1 + (-1 + i)c_2 + c_4 = 0$$
$$ic_3 + c_4 = 0$$
$$-c_3 + ic_4 = 0.$$

Set $c_4 = 1$, which makes $c_3 = i$. Thus, it is necessary to solve

$$(-1 + i)c_1 + c_2 + 1 = 0$$
$$-c_1 + (-1 + i)c_2 + 1 = 0$$

to obtain

$$c_1 = \frac{i-2}{2i-1} = \frac{4+3i}{5}$$

$$c_2 = \frac{i}{2i-1} = \frac{2-i}{5}.$$

Thus, an eigenvector is given by

$$\begin{bmatrix} \dfrac{4+3i}{5} \\ \dfrac{2-i}{5} \\ i \\ 1 \end{bmatrix}$$

and a solution takes the form

$$\begin{bmatrix} \dfrac{4+3i}{5} \\ \dfrac{2-i}{5} \\ i \\ 1 \end{bmatrix} e^{2t}[\cos(t) - \sin(t)].$$

Collecting real and imaginary parts produces two solution vectors

$$\begin{bmatrix} \frac{1}{5}(4\cos(t) + 3\sin(t)) \\ \frac{1}{5}(2\cos(t) - \sin(t)) \\ \sin(t) \\ \cos(t) \end{bmatrix} e^{2t}, \qquad \begin{bmatrix} \frac{1}{5}(3\cos(t) - 4\sin(t)) \\ \frac{1}{5}(-\cos(t) - 2\sin(t)) \\ \cos(t) \\ -\sin(t) \end{bmatrix} e^{2t}.$$

These four real solutions then form the columns of the fundamental matrix

$$\begin{bmatrix} e^t \sin(t) & e^t \cos(t) & e^{2t}\frac{1}{5}(4\cos(t) + 3\sin(t)) & e^{2t}\frac{1}{5}(3\cos(t) - 4\sin(t)) \\ e^t \cos(t) & -e^t \sin(t) & e^{2t}\frac{1}{5}(2\cos(t) - \sin(t)) & e^{2t}\frac{1}{5}(-\cos(t) - 2\sin(t)) \\ 0 & 0 & e^{2t}\sin(t) & e^{2t}\cos(t) \\ 0 & 0 & e^{2t}\cos(t) & -e^{2t}\sin(t) \end{bmatrix}.$$

EXERCISES

1. Apply the definition of derivative to the function

$$f(t) = u(t) + iv(t)$$

to show that

$$f'(t) = u'(t) + iv'(t).$$

2. Given Euler's formula, $e^{i\theta} = \cos(\theta) + i\sin(\theta)$, use Exercise 1 to show that

$$(e^{it})' = ie^{it}.$$

3. If λ is a complex number and n is an integer, show that $\overline{\lambda^n} = \overline{\lambda}^n$. If ω is also a complex number, show that $\overline{\omega\lambda} = \overline{\omega}\overline{\lambda}$.

4. Find the eigenvalues of the following matrices:

   (a) $\begin{pmatrix} 1 & 1 \\ -1 & 1 \end{pmatrix}$ (b) $\begin{pmatrix} 2 & -3 \\ 3 & 2 \end{pmatrix}$ (c) $\begin{pmatrix} 0 & 1 \\ -1 & 0 \end{pmatrix}$

5. Find a fundamental matrix for $x' = Ax$ where A is as given in Exercise 4.

6. Show that the vectors (6.3) and (6.4) are linearly indepenent.

7. Find the eigenvalues of the following matrices.

   (a) $\begin{pmatrix} 0 & 0 & 1 \\ -1 & 1 & 1 \\ -3 & 2 & 0 \end{pmatrix}$ (b) $\begin{pmatrix} 2 & 0 & 0 \\ 0 & 1 & -1 \\ 1 & 1 & 1 \end{pmatrix}$

   (c) $\begin{pmatrix} -1 & 1 & 1 \\ -3 & 3 & 1 \\ -4 & 3 & 0 \end{pmatrix}$ (d) $\begin{pmatrix} 2 & -1 & 0 & 0 \\ 1 & 2 & 0 & 0 \\ 0 & 0 & 1 & 2 \\ 0 & 0 & -2 & 1 \end{pmatrix}$

8. Find eigenvectors corresponding to the eigenvalues found in Exercise 7.

9. Find a fundamental matrix for $x' = Ax$ for each A given in Exercise 7.

10. (a) Derive the Taylor series expansion for $f(\theta) = e^{i\theta}$. (Proceed exactly as you would for real functions, using $d/d\theta(e^{i\theta}) = ie^{i\theta}$.)

    (b) Rearrange the series in (a) into real and purely imaginary parts (each part will be a series).

    (c) Identify the series in (b) and deduce Euler's formula,

    $$e^{i\theta} = \cos(\theta) + i\sin(\theta).$$

    (d) What do you need to know about the convergence of a series to perform the rearrangement in (b)?

7. The Constant Coefficient Case: The Putzer Algorithm

The analysis of the preceding section depended on finding either n distinct eigenvalues or sufficient linearly independent eigenvectors when an eigenvalue corresponded to a repeated root of the characteristic polynomial. This is the case whenever the coefficient matrix A is similar to a diagonal matrix—whenever A is "diagonalizable." When the matrix A does not have this property, the computation of e^{At} becomes difficult. To continue the approach that we have begun would require the introduction of more sophisticated linear algebra, not covered in the usual elementary course (the Jordan canonical form). For this reason we abandon the present approach and turn instead to the *Putzer algorithm*, a method for computing e^{At} based on a relatively simple theorem, the *Cayley-Hamilton theorem*, which is traditionally a part of elementary courses in linear algebra.

Let $p(\lambda)$ be a polynomial, $p(\lambda) = a_0 \lambda^n + a_1 \lambda^{n-1} + \cdots + a_n$. Since powers of square matrices make sense, we can write a corresponding matrix polynomial,

$$p(A) = a_0 A^n + a_1 A^{n-1} + \cdots + a_n I$$

(where, as above, the a_i's are scalars). The partial sums used in defining e^A were such polynomials. For every choice of a matrix A, $p(A)$ is a well-defined matrix.

THEOREM   *(Cayley-Hamilton)*
Let A be an $n \times n$ matrix and let $p(\lambda) = \det(A - \lambda I)$. Then $p(A) = 0$.

The zero, of course, is the $n \times n$ null matrix. Armed with only this theorem, we can establish the following:

THEOREM 7.1   *(Putzer)*
Let A be an $n \times n$ matrix with eigenvalues $\lambda_1, \lambda_2, \ldots, \lambda_n$. Then

$$e^{At} = \sum_{j=0}^{n-1} r_{j+1}(t) P_j \tag{7.1}$$

where $P_0 = I$,

$$P_j = \prod_{k=1}^{j} (A - \lambda_k I), \qquad j = 1, \ldots, n, \tag{7.2}$$

and $r_1(t), \ldots, r_n(t)$ is the solution of the triangular system

$$r_1' = \lambda_1 r_1$$

$$r_j' = r_{j-1} + \lambda_j r_j, \qquad j = 2, \ldots, n, \tag{7.3}$$

$$r_1(0) = 1, \qquad r_j(0) = 0, \qquad j = 2, \ldots, n.$$

Note first that each eigenvalue appears in the list repeated according to its multiplicity. Further, note that the order of the matrices is not crucial—$A - \lambda_1 I$ and $A - \lambda_2 I$ commute—so for convenience in the computation, we adopt the convention that $(A - \lambda_i I)$ precedes $(A - \lambda_j I)$ if $i > j$ in the product. The system (7.3) can be solved recursively; if $r_1(t)$ is found, the equation for $r_2(t)$ is a *forced* first-order linear equation, the "forcing" being $r_1(t)$. This process can be continued until $r_1(t), \ldots, r_n(t)$ are found merely by solving first-order linear differential equations.

Proof. Let $\Phi(t) = \sum_{j=0}^{n-1} r_{j+1}(t) P_j$. The idea of the proof is to show that $\Phi(t)$ satisfies $\Phi' = A\Phi$, $\Phi(0) = I$ so that $\Phi(t) = e^{At}$, by the uniqueness of solutions. For convenience, define $r_0(t) \equiv 0$. Then

$$\Phi'(t) = \sum_{j=0}^{n-1} r_{j+1}'(t) P_j$$

$$= \sum_{j=0}^{n-1} [\lambda_{j+1} r_{j+1}(t) + r_j(t)] P_j$$

so that

$$\Phi'(t) - \lambda_n \Phi = \sum_{j=0}^{n-1} (\lambda_{j+1} r_{j+1}(t) + r_j(t)) P_j - \lambda_n \sum_{j=0}^{n-1} r_{j+1}(t) P_j$$

$$= \sum_{j=0}^{n-1} (\lambda_{j+1} - \lambda_n) r_{j+1}(t) P_j + \sum_{j=0}^{n-1} r_j(t) P_j$$

$$= \sum_{j=0}^{n-2} (\lambda_{j+1} - \lambda_n) r_{j+1}(t) P_j + \sum_{j=0}^{n-2} r_{j+1}(t) P_{j+1}.$$

Since $P_{j+1} = (A - \lambda_{j+1} I) P_j$ by (7.2), the last line may be rewritten as

$$\Phi'(t) - \lambda_n \Phi(t) = \sum_{j=0}^{n-2} [(A - \lambda_{j+1} I) P_j + (\lambda_{j+1} - \lambda_n) P_j] r_{j+1}(t)$$

$$= \sum_{j=0}^{n-2} (A - \lambda_n I) P_j r_{j+1}(t)$$

$$= (A - \lambda_n I) \sum_{j=0}^{n-2} P_j r_{j+1}(t).$$

We manipulate this right-hand side so as to obtain the appropriate equation for Φ. Since

$$\sum_{j=0}^{n-2} P_j r_{j+1} = \sum_{j=0}^{n-1} P_j r_{j+1}(t) - P_{n-1} r_n(t)$$

$$= \Phi(t) - r_n(t) P_{n-1},$$

then

$$\Phi'(t) - \lambda_n \Phi(t) = (A - \lambda_n I)\Phi(t) - r_n(t)(A - \lambda_n I)P_{n-1}$$

$$= (A - \lambda_n I)\Phi(t) - r_n(t) P_n.$$

The characteristic equation for A may be written in factored form as

$$p(\lambda) = (\lambda - \lambda_n)(\lambda - \lambda_{n-1}) \cdots (\lambda - \lambda_2)(\lambda - \lambda_1).$$

Since

$$P_n = (A - \lambda_n I)(A - \lambda_{n-1} I) \cdots (A - \lambda_1 I)$$

$$= p(A),$$

it follows by the Cayley–Hamilton theorem that $P_n = 0$ (the null matrix). Therefore, $\Phi(t)$ satisfies the differential equation

$$\Phi'(t) = A\Phi(t) \tag{7.4}$$

and the initial condition

$$\Phi(0) = \sum_{j=0}^{n-1} r_{j+1}(0) P_j = r_1(0) I$$

$$= I.$$

Hence, it follows by the uniqueness of solutions of (7.4) that $\Phi(t) = e^{At}$.

We illustrate the theorem first for a simple two-dimensional system. Consider

$$y' = \begin{pmatrix} 3 & -1 \\ 1 & 1 \end{pmatrix} y.$$

First, solve

$$\det(A - \lambda I) = \det \begin{pmatrix} 3 - \lambda & -1 \\ 1 & 1 - \lambda \end{pmatrix}$$

$$= 3 - 4\lambda + \lambda^2 + 1 = 0$$

and find that $\lambda = 2$ is a double root. Following the algorithm, let $\lambda_1 = 2$, $\lambda_2 = 2$,

$P_0 = I$, and

$$P_1 = (A - 2I) = \begin{pmatrix} 1 & -1 \\ 1 & -1 \end{pmatrix}.$$

Further,

$$r_1' = 2r_1$$
$$r_1(0) = 1$$

so that $r_1(t) = e^{2t}$. Since

$$r_2' = e^{2t} + 2r_2$$
$$r_2(0) = 0,$$

we have $r_2(t) = te^{2t}$. Therefore,

$$e^{At} = e^{2t}\begin{pmatrix} 1 & 0 \\ 0 & 1 \end{pmatrix} + te^{2t}\begin{pmatrix} 1 & -1 \\ 1 & -1 \end{pmatrix}$$

$$= e^{2t}\begin{pmatrix} 1+t & -t \\ t & 1-t \end{pmatrix}.$$

Consider now the system

$$y' = \begin{pmatrix} 1 & 1 & 0 \\ 0 & 1 & 0 \\ -1 & 2 & 2 \end{pmatrix} y.$$

The characteristic polynomial, $\det(A - \lambda I) = 0$, takes the form

$$(1 - \lambda)^2(2 - \lambda) = 0$$

and we label the roots $\lambda_1 = \lambda_2 = 1$, $\lambda_3 = 2$. Then $P_0 = I$,

$$P_1 = \begin{pmatrix} 0 & 1 & 0 \\ 0 & 0 & 0 \\ -1 & 2 & 1 \end{pmatrix},$$

and

$$P_2 = P_1^2 = \begin{pmatrix} 0 & 0 & 0 \\ 0 & 0 & 0 \\ -1 & 1 & 1 \end{pmatrix}.$$

Now solve the system (7.3) recursively. From

$$r_1' = r_1$$
$$r_1(0) = 1$$

it follows that

$$r_1(t) = e^t.$$

From

$$r_2' = e^t + r_2$$
$$r_2(0) = 0$$

it follows that

$$r_2(t) = te^t.$$

Finally, from

$$r_3' = te^t + 2r_3$$
$$r_3(0) = 0$$

it follows that

$$r_3(t) = e^{2t} - te^t - e^t.$$

Thus, from the Putzer algorithm we have that

$$e^{At} = e^t \begin{pmatrix} 1 & 0 & 0 \\ 0 & 1 & 0 \\ 0 & 0 & 1 \end{pmatrix} + te^t \begin{pmatrix} 0 & 1 & 0 \\ 0 & 0 & 0 \\ -1 & 2 & 1 \end{pmatrix} + (e^{2t} - te^t - e^t) \begin{pmatrix} 0 & 0 & 0 \\ 0 & 0 & 0 \\ -1 & 1 & 1 \end{pmatrix}$$

or

$$e^{At} = \begin{pmatrix} e^t & te^t & 0 \\ 0 & e^t & 0 \\ -e^{2t} + e^t & e^{2t} + te^t - e^t & e^{2t} \end{pmatrix}.$$

As a final example, consider

$$y' = \begin{pmatrix} 2 & 1 & 0 & 0 \\ 0 & 2 & 1 & 0 \\ 0 & 0 & 2 & 0 \\ 0 & 0 & 0 & 3 \end{pmatrix} y.$$

Then $\det(A - \lambda I) = 0$ yields $(2 - \lambda)^3 (3 - \lambda) = 0$ and we label the roots $\lambda_1 = \lambda_2 = \lambda_3 = 2$, $\lambda_4 = 3$. Then $P_0 = I$,

$$P_1 = \begin{pmatrix} 0 & 1 & 0 & 0 \\ 0 & 0 & 1 & 0 \\ 0 & 0 & 0 & 0 \\ 0 & 0 & 0 & 1 \end{pmatrix},$$

$$P_2 = P_1^2 = \begin{pmatrix} 0 & 0 & 1 & 0 \\ 0 & 0 & 0 & 0 \\ 0 & 0 & 0 & 0 \\ 0 & 0 & 0 & 1 \end{pmatrix},$$

and

$$P_3 = P_1^3 = \begin{pmatrix} 0 & 0 & 0 & 0 \\ 0 & 0 & 0 & 0 \\ 0 & 0 & 0 & 0 \\ 0 & 0 & 0 & 1 \end{pmatrix}.$$

We proceed to find the functions $r_i(t)$, $i = 1, 2, 3, 4$. First of all,

$$r_1' = 2r_1$$
$$r_1(0) = 1$$

so $r_1(t) = e^{2t}$. Then, $r_2(t)$ satisfies

$$r_2' = e^{2t} + 2r_2$$
$$r_2(0) = 0$$

or

$$r_2(t) = te^{2t},$$

and $r_3(t)$ satisfies

$$r_3' = te^{2t} + 2r_3$$
$$r_3(0) = 0$$

or

$$r_3(t) = \frac{t^2}{2}e^{2t}.$$

Finally, $r_4(t)$ satisfies

$$r_4' = \frac{t^2}{2}e^{2t} + 3r_4$$

$$r_4(0) = 0$$

or

$$e^{-3t}r_4(t) = -\frac{t^2 e^{-t}}{2} - te^{-t} - e^{-t} + 1$$

or

$$r_4(t) = e^{3t} - e^{2t} - te^{2t} - \frac{t^2}{2}e^{2t}.$$

Thus

$$e^{At} = e^{2t}\begin{pmatrix} 1 & 0 & 0 & 0 \\ 0 & 1 & 0 & 0 \\ 0 & 0 & 1 & 0 \\ 0 & 0 & 0 & 1 \end{pmatrix} + te^{2t}\begin{pmatrix} 0 & 1 & 0 & 0 \\ 0 & 0 & 1 & 0 \\ 0 & 0 & 0 & 0 \\ 0 & 0 & 0 & 1 \end{pmatrix} + \frac{t^2}{2}e^{2t}\begin{pmatrix} 0 & 0 & 1 & 0 \\ 0 & 0 & 0 & 0 \\ 0 & 0 & 0 & 0 \\ 0 & 0 & 0 & 1 \end{pmatrix}$$

$$+ \left(e^{3t} - e^{2t} - te^{2t} - \frac{t^2 e^{2t}}{2}\right)\begin{pmatrix} 0 & 0 & 0 & 0 \\ 0 & 0 & 0 & 0 \\ 0 & 0 & 0 & 0 \\ 0 & 0 & 0 & 1 \end{pmatrix}$$

or

$$e^{At} = \begin{pmatrix} e^{2t} & te^{2t} & \dfrac{t^2}{2}e^{2t} & 0 \\ 0 & e^{2t} & te^{2t} & 0 \\ 0 & 0 & e^{2t} & 0 \\ 0 & 0 & 0 & e^{3t} \end{pmatrix}.$$

All of the illustrations above were for the case of a repeated root, since this is the case where the previous method could fail. However, the method works equally well in the case of distinct roots. It finds e^{At} directly and avoids finding the inverse of an arbitrary fundamental matrix. (Recall that $e^{At} = \Phi(t)\Phi^{-1}(0)$ where $\Phi(t)$ is an arbitrary fundamental matrix.) The computations with the Putzer algorithm are usually more involved than computing the required eigenvectors for a fundamental matrix.

In the illustrations, the eigenvalues were all real. However, the method works equally well if they are complex, since the differential equations for the functions r can be solved in just the same manner. For example, the equation

$$y' + iy = 0$$

has the general solution $y(t) = ce^{-it}$, where c is constant. Solutions with real initial conditions are no longer necessarily real, but otherwise everything is as before. For example, if we add the initial condition $y(0) = 1$, then the solution is $y(t) = e^{-it}$. We illustrate the Putzer algorithm with a simple example.

Consider the system

$$z' = \begin{pmatrix} 0 & 1 & 0 & 0 \\ -1 & 0 & 0 & 0 \\ 0 & 0 & 0 & 1 \\ 0 & 0 & -1 & 0 \end{pmatrix} z.$$

The eigenvalues are roots of the polynomial

$$p(\lambda) = \det \begin{pmatrix} -\lambda & 1 & 0 & 0 \\ -1 & -\lambda & 0 & 0 \\ 0 & 0 & -\lambda & 1 \\ 0 & 0 & -1 & -\lambda \end{pmatrix} = 0.$$

Expanding the determinant yields

$$p(\lambda) = \lambda^2(\lambda^2 + 1) + (\lambda^2 + 1)$$
$$= (\lambda^2 + 1)^2 = 0$$

or that

$$\lambda = \pm i$$

are double roots. To apply the algorithm, label

$$\lambda_1 = i, \quad \lambda_2 = -i,$$
$$\lambda_3 = i, \quad \lambda_4 = -i.$$

First of all,

$$P_0 = I,$$

$$P_1 = \begin{pmatrix} -i & 1 & 0 & 0 \\ -1 & -i & 0 & 0 \\ 0 & 0 & -i & 1 \\ 0 & 0 & -1 & -i \end{pmatrix},$$

$$P_2 = (A + iI)P_1 = 0$$

(the null matrix), and hence

$$P_3 = 0.$$

Also, $r_1(t)$ satisfies the equation

$$r_1' = ir_1, \qquad r_1(0) = 1,$$

so $r_1(t) = e^{it}$, and $r_2(t)$ satisfies

$$r_2' = -ir_2 + e^{it}.$$

A straightforward computation shows that

$$r_2(t) = \frac{1}{2i}[e^{it} - e^{-it}]$$

$$= \sin(t).$$

Since P_2 and P_3 are null, there is no need to compute $r_3(t)$ and $r_4(t)$. The solution is real even if the differential equation is complex! Since $e^{At} = e^{it}I + \sin(t)P_1$, we have

$$
\begin{pmatrix}
e^{it} - i\sin(t) & \sin(t) & 0 & 0 \\
-\sin(t) & e^{it} - i\sin(t) & 0 & 0 \\
0 & 0 & e^{it} - i\sin(t) & \sin(t) \\
0 & 0 & -\sin(t) & e^{it} - i\sin(t)
\end{pmatrix}.
$$

However, $e^{it} - i\sin(t) = \cos(t) + i\sin(t) - i\sin(t) = \cos(t)$. Thus, e^{At} is equal to

$$
\begin{pmatrix}
\cos(t) & \sin(t) & 0 & 0 \\
-\sin(t) & \cos(t) & 0 & 0 \\
0 & 0 & \cos(t) & \sin(t) \\
0 & 0 & -\sin(t) & \cos(t)
\end{pmatrix}.
$$

Of course, e^{At} had to turn out to be real, since A was real. The algorithm simply took us through an excursion in the complex domain. The result is so simple because we were, in effect, dealing with two uncoupled systems. A more interesting computation occurs if we change the system slightly to

$$
z' = \begin{pmatrix}
0 & 1 & 0 & 1 \\
-1 & 0 & 0 & 0 \\
0 & 0 & 0 & 1 \\
0 & 0 & -1 & 0
\end{pmatrix} z.
$$

The eigenvalues are roots of the polynomial

$$
p(\lambda) = \det \begin{pmatrix}
-\lambda & 1 & 0 & 1 \\
-1 & -\lambda & 0 & 0 \\
0 & 0 & -\lambda & 1 \\
0 & 0 & -1 & -\lambda
\end{pmatrix} = 0.
$$

Expanding the determinant yields the same polynomial as in the previous example,

$$
p(\lambda) = (\lambda^2 + 1)^2 = 0,
$$

or that

$$
\lambda = \pm i
$$

are again double roots. Take the same definition of the order of the eigenvalues, $\lambda_1 = i$, $\lambda_2 = -i$, $\lambda_3 = i$, $\lambda_4 = -i$, and proceed with the computation. As always, $P_0 = I$ and

$$P_1 = \begin{pmatrix} -i & 1 & 0 & 1 \\ -1 & -i & 0 & 0 \\ 0 & 0 & -i & 1 \\ 0 & 0 & -1 & -i \end{pmatrix},$$

$$P_2 = \begin{pmatrix} 0 & 0 & -1 & 0 \\ 0 & 0 & 0 & -1 \\ 0 & 0 & 0 & 0 \\ 0 & 0 & 0 & 0 \end{pmatrix},$$

and

$$P_3 = \begin{pmatrix} 0 & 0 & i & -1 \\ 0 & 0 & 1 & i \\ 0 & 0 & 0 & 0 \\ 0 & 0 & 0 & 0 \end{pmatrix}.$$

We find $r_1(t)$ and $r_2(t)$ as before. To find $r_3(t)$, we must solve

$$r_3' = ir_3 + \sin(t).$$

If this quantity is rewritten as

$$(e^{-it}r_3(t))' = e^{-it}\sin(t)$$

we see that the solution with $r_3(0) = 0$ is given by

$$r_3(t) = \tfrac{1}{2}[i\sin(t) - it\cos(t) + t\sin(t)].$$

Finally, r_4 is the solution of

$$r_4' = -ir_4 + r_3$$

and can be obtained by integrating both sides of

$$(r_4 e^{it})' = e^{it}r_3.$$

An integration and some manipulation yields that

$r_4(t) = \frac{1}{2}\sin(t) - t\cos(t).$

Thus, the matrix e^{At} takes the form

$$
\begin{pmatrix}
\cos(t) & \sin(t) & -\dfrac{t}{2}\sin(t) & \dfrac{1}{2}(t\cos(t) + \sin(t)) \\
-\sin(t) & \cos(t) & \frac{1}{2}(\sin(t) - x\cos(t)) & \frac{1}{2}t\sin(t) \\
0 & 0 & \cos(t) & \sin(t) \\
0 & 0 & -\sin(t) & \cos(t)
\end{pmatrix}.
$$

As an aside, for those interested in computing, this algorithm can easily be set up to be performed by a symbolic manipulator, such as *Reduce*.

EXERCISES

1. Find e^{At} where

(a) $A = \begin{pmatrix} 1 & 1 \\ 0 & 1 \end{pmatrix}$

(b) $A = \begin{pmatrix} 2 & 6 \\ -2 & -4 \end{pmatrix}$

(c) $A = \begin{pmatrix} 2 & 1 \\ -1 & 4 \end{pmatrix}$

2. Find e^{At} where

(a) $A = \begin{pmatrix} 1 & 1 & 1 \\ 2 & 1 & -1 \\ -3 & 2 & 0 \end{pmatrix}$

(b) $A = \begin{pmatrix} 2 & 1 & 0 \\ 0 & 2 & 0 \\ 0 & 1 & 2 \end{pmatrix}$

(c) $A = \begin{pmatrix} 2 & -2 & 0 \\ 0 & 2 & 0 \\ -2 & 2 & 2 \end{pmatrix}$

(d) $A = \begin{pmatrix} 6 & -2 & 1 \\ 5 & 0 & 1 \\ 2 & -1 & 3 \end{pmatrix}$

(e) $A = \begin{pmatrix} 1 & 0 & 1 \\ -1 & 2 & 1 \\ -1 & 1 & 1 \end{pmatrix}$

3. Find the solution of $x' = Ax$, $x(0) = \begin{pmatrix} 1 \\ -1 \\ 1 \end{pmatrix}$ for each A given in Exercise 2.

4. Verify the Cayley-Hamilton theorem for each of the matrices in Exercise 2.

5. Let $A = \begin{pmatrix} B & 0 \\ 0 & C \end{pmatrix}$ where B is an $r \times r$ matrix, C is a $p \times p$ matrix, and A is an $(n + p) \times (n + p)$ matrix. Show that

$$e^{At} = \begin{pmatrix} e^{Bt} & 0 \\ 0 & e^{Ct} \end{pmatrix} \quad \text{and if } A \text{ is invertible, } A^{-1} = \begin{pmatrix} B^{-1} & 0 \\ 0 & C^{-1} \end{pmatrix}.$$

6. Use Exercise 5 to find e^{At} where $A = \begin{pmatrix} 1 & 1 & 0 & 0 \\ 0 & 1 & 0 & 0 \\ 0 & 0 & 2 & 1 \\ 0 & 0 & 0 & 2 \end{pmatrix}$.

7. Show that if the real part of each eigenvalue of A is negative, then every solution of $y' = Ay$ satisfies $\lim_{t \to \infty} y(t) = 0$. (*Hint:* Show that $\lim_{t \to \infty} r_i(t) = 0$.)

8. The following matrices have complex eigenvalues. Use the Putzer algorithm to find e^{At}.

(a) $\begin{pmatrix} 2 & 1 \\ -1 & 2 \end{pmatrix}$

(b) $\begin{pmatrix} 1 & 2 & 1 \\ -2 & 1 & 0 \\ 0 & 0 & 3 \end{pmatrix}$

(c) $\begin{pmatrix} 0 & 1 & 0 & 0 \\ -1 & 0 & 0 & 0 \\ 0 & 0 & 0 & 1 \\ 1 & 0 & -1 & 0 \end{pmatrix}$

8. General Linear Systems

We now consider a linear system with a *forcing term*,

$$y' = A(t)y + e(t), \tag{8.1}$$

where $A(t)$ is an $n \times n$ continuous matrix and $e(t)$ is a continuous n-vector. For notational purposes, let $L[y]$ denote $y' - Ay$. Then, as noted before, (8.1) can be written

$$L[y] = e. \tag{8.2}$$

The principal theoretical result is given in the following theorem. Note first that L is a linear operator.

THEOREM 8.1

   If $\chi(t)$ is a given solution of (8.2), then any solution $\psi(t)$ of (8.2) can be written

$$\psi(t) = \Phi(t)c + \chi(t)$$

where Φ is a fundamental matrix for

$$L[y] = 0 \tag{8.3}$$

and c is an appropriate constant vector.

Proof.   Let Φ be a fundamental matrix for $L[y] = 0$, and let $\psi(t)$ be an arbitrary solution. Then

$$L[\psi - \chi] = L[\psi] - L[\chi]$$
$$= e - e = 0$$

or $\psi(t) - \chi(t)$ is a solution of (8.3). By Theorem 3.4,

$$\psi(t) - \chi(t) = \Phi(t)c,$$

or

$$\psi(t) = \Phi(t)c + \chi(t),$$

which is the conclusion of the theorem.

The importance of Theorem 8.1 is that it reduces the problem of finding all solutions of equation (8.1) to the problem of finding a fundamental matrix for (8.3) and finding any solution whatever of (8.1). We deal now with a way to find $\chi(t)$ given the fundamental matrix $\Phi(t)$.

THEOREM 8.2
The vector function

$$\chi(t) = \Phi(t) \int_{t_0}^{t} \Phi^{-1}(\tau)e(\tau) \, d\tau \tag{8.4}$$

is a solution of (8.1).

Proof.   To prove the theorem it is necessary only to verify by differentiation that (8.4) is a solution. To check first, however, that the above makes sense, note that Φ and Φ^{-1} are $n \times n$ matrices, so $\Phi^{-1}(\tau)e(\tau)$ is an n vector, as is its integral, and $\Phi(t)$ operates on the n vector $\int_{t_0}^{t} \Phi^{-1}(\tau)e(\tau) \, d\tau$. Of course, $\Phi(t)$ is differentiable, and since $e(t)$ is continuous, so is the integral in (8.4). Differentiating,

$$\chi'(t) = \Phi'(t) \int_{t_0}^{t} \Phi^{-1}(\tau)e(\tau) \, d\tau + \Phi(t)\Phi^{-1}(t)e(t).$$

Since $\Phi'(t) = A(t)\Phi(t)$, and since $\Phi^{-1}(t)\Phi(t) = I$, this becomes

$$\chi'(t) = A(t)\Phi(t) \int_{t_0}^{t} \Phi^{-1}(\tau)e(\tau) \, d\tau + e(t).$$

By the definition of $\chi(t)$ (8.4), this is

$$\chi'(t) = A(t)\chi(t) + e(t)$$

or

$$L[\chi] = e.$$

Note that $\chi(t_0) = 0$; that is, $\chi(t)$ is the solution of (8.1) that takes the zero vector as the initial condition at $t = t_0$. Suppose we desire to solve (8.1) with initial conditions $y(t_0) = \eta$. Let $\Phi(t)$ be a fundamental matrix for (8.3). If $\chi(t)$ is given by (8.4), then

$$\psi(t) = \Phi(t)\Phi^{-1}(t_0)\eta + \chi(t)$$

is a solution of the equation, for

$$
\begin{aligned}
\psi'(t) &= \Phi'(t)\Phi^{-1}(t_0)\eta + \chi'(t) \\
&= A(t)\Phi(t)\Phi^{-1}(t_0)\eta + A(t)\chi(t) + e(t) \\
&= A(t)(\Phi(t)\Phi^{-1}(t_0)\eta + \chi(t)) + e(t) \\
&= A(t)\psi(t) + e(t).
\end{aligned}
$$

Further,

$$
\begin{aligned}
\Psi(t_0) &= \Phi(t_0)\Phi^{-1}(t_0)\eta + \chi(t_0) \\
&= I\eta = \eta,
\end{aligned}
$$

so $\psi(t)$ satisfies the initial condition.

This computation can be combined with Theorems 8.1 and 8.2 to yield a solution for any linear initial value problem.

THEOREM 8.3 *(Variation of constants formula)*
Let $\Phi(t)$ be a fundamental matrix for $\chi' = A(t)\chi$. Then the unique solution of

$$y' = A(t)y + e(t)$$

$$y(t_0) = \eta$$

is given by

$$y(t) = \Phi(t)\Phi^{-1}(t_0)\eta + \int_{t_0}^{t} \Phi(t)\Phi^{-1}(s)e(s)\,ds. \qquad (8.5)$$

In the representation of the solution, (8.5), the matrix $\Phi(t)$ has been moved inside the integral sign. This causes no difficulties, since the integration is with respect to the dummy variable s. However, it does make for a nice formula if the fundamental matrix $\Phi(t)$ happens to be e^{At}. Then with $t_0 = 0$ the representation given in Theorem 8.3 is

$$y(t) = e^{At}\eta + \int_0^t e^{A(t-s)}e(s)\,ds. \tag{8.6}$$

Equation (8.6) follows from (8.5), since $e^{A0} = I$ and $(e^{At})^{-1} = e^{-At}$. As an example, consider the system

$$\begin{aligned} y_1' &= y_2 \\ y_2' &= -y_1 + t, \end{aligned} \tag{8.7}$$

and the initial conditions

$$\begin{aligned} y_1(0) &= 0 \\ y_2(0) &= 2, \end{aligned} \tag{8.8}$$

or

$$y' = Ay + e(t)$$

where

$$A = \begin{bmatrix} 0 & 1 \\ -1 & 0 \end{bmatrix}$$

and

$$e(t) = \begin{bmatrix} 0 \\ t \end{bmatrix}.$$

The unforced equation is

$$\begin{aligned} y_1' &= y_2 \\ y_2' &= -y_1. \end{aligned} \tag{8.9}$$

A fundamental matrix for (8.10) is

$$\Phi(t) = \begin{bmatrix} \cos(t) & \sin(t) \\ -\sin(t) & \cos(t) \end{bmatrix} \tag{8.10}$$

because the columns of (8.10) are solutions of (8.9) and $\det \Phi(t) = \cos^2(t) + \sin^2(t) = 1$. Then,

$$\Phi^{-1}(t) = \begin{bmatrix} \cos(t) & -\sin(t) \\ \sin(t) & \cos(t) \end{bmatrix}.$$

Hence, $\chi(t)$ is given by

$$\chi(t) = \Phi(t) \int_0^t \Phi^{-1}(\tau) e(\tau) \, d\tau$$

$$= \begin{bmatrix} \cos(t) & \sin(t) \\ -\sin(t) & \cos(t) \end{bmatrix} \begin{bmatrix} -\int_{t_0}^t \tau \sin(\tau) \, d\tau \\ \int_{t_0}^t \tau \cos(\tau) \, d\tau \end{bmatrix}$$

$$= \begin{bmatrix} \cos(t) & \sin(t) \\ -\sin(t) & \cos(t) \end{bmatrix} \begin{bmatrix} -\sin(t) + t\cos(t) \\ \cos(t) + t\sin(t) - 1 \end{bmatrix}$$

$$= \begin{bmatrix} -\sin(t)\cos(t) + t\cos^2(t) + \sin(t)\cos(t) + t\sin^2(t) - \sin(t) \\ +\sin^2(t) - t\sin(t)\cos(t) + \cos^2(t) + t\sin(t)\cos(t) - \cos(t) \end{bmatrix}$$

$$= \begin{bmatrix} t - \sin(t) \\ 1 - \cos(t) \end{bmatrix}.$$

To satisfy the initial conditions, (8.8), we have

$$c = \Phi^{-1}(0) \begin{bmatrix} 0 \\ 2 \end{bmatrix}$$

or

$$c = \begin{bmatrix} 1 & 0 \\ 0 & 1 \end{bmatrix} \begin{bmatrix} 0 \\ 2 \end{bmatrix} = \begin{bmatrix} 0 \\ 2 \end{bmatrix}.$$

Finally,

$$\psi(t) = \Phi(t)c + \chi(t)$$

$$= \begin{bmatrix} \cos(t) & +\sin(t) \\ -\sin(t) & \cos(t) \end{bmatrix} \begin{bmatrix} 0 \\ 2 \end{bmatrix} + \begin{bmatrix} t - \sin(t) \\ 1 - \cos(t) \end{bmatrix}$$

$$= \begin{bmatrix} 2\sin(t) \\ 2\cos(t) \end{bmatrix} + \begin{bmatrix} t - \sin(t) \\ 1 - \cos(t) \end{bmatrix}$$

$$= \begin{bmatrix} t + \sin(t) \\ 1 + \cos(t) \end{bmatrix}.$$

Consider the 3×3 system

$$y' = \begin{bmatrix} 2 & 1 & 1 \\ 0 & 2 & 0 \\ 0 & 0 & 3 \end{bmatrix} y + \begin{bmatrix} 1 \\ 0 \\ t \end{bmatrix} \tag{8.11}$$

with the initial condition

$$y(0) = \begin{bmatrix} 1 \\ 1 \\ 1 \end{bmatrix}. \tag{8.12}$$

Let A denote the above 3×3 matrix. The eigenvalues of A are given by roots of

$$p(\lambda) = \det \begin{bmatrix} 2 - \lambda & 1 & 1 \\ 0 & 2 - \lambda & 0 \\ 0 & 0 & 3 - \lambda \end{bmatrix} = 0.$$

Expanding the determinant, we see that

$$p(\lambda) = (2 - \lambda)^2 (3 - \lambda) = 0$$

or that $\lambda_1 = 2, \lambda_2 = 2$, and $\lambda_3 = 3$. We apply the Putzer algorithm to find e^{At}. The relevant matrices are $P_0 = I$,

$$P_1 = \begin{bmatrix} 0 & 1 & 1 \\ 0 & 0 & 0 \\ 0 & 0 & 1 \end{bmatrix},$$

and

$$P_2 = \begin{bmatrix} 0 & 0 & 1 \\ 0 & 0 & 0 \\ 0 & 0 & 1 \end{bmatrix}.$$

Clearly, $r_1(t) = e^{2t}$ and $r_2(t)$ solves

$$r_2' = 2r_2 + e^{2t}, \qquad r_2(0) = 0,$$

or $r_2(t) = te^{2t}$. Then $r_3(t)$ satisfies

$$r_3' = 3r_3 + te^{2t}, \qquad r_3(0) = 0.$$

Rewriting the differential equation as

$$(r_3 e^{-3t})' = te^{-t}$$

gives, after an integration, $r_3(t) = e^{2t}(e^t - t - 1)$. Thus, e^{At} takes the form

$$\begin{bmatrix} e^{2t} & te^{2t} & e^{3t} - e^{2t} \\ 0 & e^{2t} & 0 \\ 0 & 0 & e^{3t} \end{bmatrix}. \tag{8.13}$$

Rather than directly invert this matrix, we can use the fact that $(e^{At})^{-1} = e^{-At}$. This amounts to substituting $-t$ for t in (8.13), which yields that Φ^{-1}, in (8.5), is of the form

$$\begin{bmatrix} e^{-2t} & -te^{-2t} & e^{-3t} - e^{-2t} \\ 0 & e^{-2t} & 0 \\ 0 & 0 & e^{-3t} \end{bmatrix}. \tag{8.14}$$

Then, $e^{-At}e(t)$ is

$$\begin{bmatrix} e^{-2t}(te^{-t} - t + 1) \\ 0 \\ te^{-3t} \end{bmatrix}.$$

We can now compute the particular solution, (8.4). An integration of the vector above from 0 to t produces

$$\begin{bmatrix} \frac{1}{36}[e^{-3t}(13e^{3t} + 18te^t - 9e^t - 12t - 4) + 13] \\ 0 \\ \frac{1}{9}e^{-3t}(e^{3t} - 3t - 1) \end{bmatrix}.$$

Multiplying e^{At} by this vector produces the χ of (8.4). Recall that this is a solution of the system that is the null vector at $t = 0$. The result of this multiplication is

$$\chi(t) = \begin{bmatrix} \frac{1}{36}[4e^{3t} + 9e^{2t} + 6t - 13] \\ 0 \\ \frac{1}{9}[e^{3t} - 3t - 1] \end{bmatrix}.$$

Finally, the variation of constants formula may be applied; it requires us to compute $e^{At}y(0) + \chi(t)$. This yields the solution of (8.11) in the form

$$y(t) = \begin{bmatrix} \frac{1}{36}[40e^{3t} + 36te^{2t} + 9e^{2t} + 6t - 13] \\ e^{2t} \\ \frac{1}{9}[10e^{3t} - 3t - 1] \end{bmatrix}.$$

While computations such as this are not intrinsically difficult, it is clear that considerable, careful manipulation is necessary to carry out the very elegant representation of the solution given by Theorem 8.3. For this reason, the variation of constants formula is of more theoretical than practical importance for all but the simplest systems. Again, if it is necessary to find the explicit representation of a large system, a computer with a symbolic manipulator is a useful tool.

EXERCISES

1. Find all solutions of

   (a) $x' = \begin{pmatrix} 1 & 2 \\ 0 & 2 \end{pmatrix} x + \begin{pmatrix} 0 \\ t \end{pmatrix}$

   (b) $x' = \begin{pmatrix} -1 & -1 \\ 2 & 4 \end{pmatrix} x + \begin{pmatrix} 1 \\ t \end{pmatrix}$

   (c) $x' = \begin{pmatrix} 2 & 2 \\ -1 & 2 \end{pmatrix} x + \begin{pmatrix} t \\ \sin(t) \end{pmatrix}$

   (d) $x' = \begin{pmatrix} 1 & 1 \\ -1 & 1 \end{pmatrix} x + \begin{pmatrix} t \\ 0 \end{pmatrix}$

2. Find the solution for each system given in Exercise 1 that satisfies $x(0) = \begin{pmatrix} 1 \\ 1 \end{pmatrix}$.

3. Determine whether the limit as $t \to \infty$ or as $t \to -\infty$ exists for any of the solutions found in Exercise 1.

4. Find all solutions of

   $$x' = \begin{pmatrix} 3 & -1 & 0 \\ 2 & 0 & 0 \\ 2 & -2 & 3 \end{pmatrix} x + \begin{pmatrix} 1 \\ t \\ t^2 \end{pmatrix}.$$

5. Find all solutions of

   $$x' = \begin{pmatrix} 1 & 1 & 0 & 0 \\ 0 & 1 & 0 & 0 \\ 0 & 0 & 0 & 1 \\ 0 & 0 & 1 & 0 \end{pmatrix} x + \begin{pmatrix} e^{-t} \\ 0 \\ e^{-t} \\ 0 \end{pmatrix}.$$

9.   Some Elementary Stability Considerations

The theory developed in the previous sections makes it possible to intro-duce some of the basic ideas of stability analysis for systems of differential equations. These ideas are important in many physical systems to give a robust-ness to theoretical conclusions. More sophisticated tools and concepts will appear in the next chapter. Although, at this point, the discussion will be restricted to linear systems (indeed, to those with constant coefficients), the concepts carry over to nonlinear systems as well. For those who have met the idea of stability in a physics course, we note that what is presented here is a mathematician's way of describing those very same ideas. The properties of norm, introduced at the beginning of Section 4, are important in this endeavor and the reader may wish to review them before proceeding with this section.

The basic, intuitive idea is that of instability. "Something" is unstable if a small deviation from the present "state" produces a major change in the state. The familiar physical example is a cone balanced on its pointed end (see Figure 9.1). A small change in position produces a major change—the cone falls. To make this, and related ideas, precise in the context of solutions of systems of linear differential equations is the goal of this section.

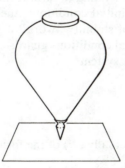

Figure 9.1   An example of instability. A cone balanced on its point will fall if slightly disturbed.

Consider the linear system of ordinary differential equations

$$x' = Ax \tag{9.1}$$

where A is an $n \times n$ constant matrix and x is a vector in R^n. Equation (9.1) always has the trivial solution, the function $x(t) = 0$, and this solution will play the role of "present state" in the intuitive description above. The trivial solution is said to be *stable* if for every $\varepsilon > 0$ there is a $\delta > 0$ such that if $x(t)$ is any solution of (9.1) with $\|x(0)\| < \delta$, then $\|x(t)\| < \varepsilon$ for all $t > 0$. We are using norm, $\| \quad \|$, to measure how close a solution is to the trivial solution. Think of the trivial solution as the present state of the system and $x(t)$ as a solution that represents a deviation from the present state. The above definition says that, if the trivial solution is stable, $x(t)$ will remain arbitrarily close (this is the ε) to the present state (the trivial solution) for all future time if the initial condition $x(0)$ is sufficiently close (this is the δ) to zero. The trivial solution is said to be *unstable* if it is not stable. Given the definition of stability, being unstable meets the above intuitive criterion of a major (not small) change of state at a future time from a small, initial disturbance (a small change of initial condition). To set the basic idea, let us redefine instability by formally contradicting the notion of stability— by stating what must happen for stability to fail. Roughly, we must state that no matter how close the initial conditions are to zero, at some future time some solution will not be close to zero. Thus it is sufficient to show that for some $\varepsilon > 0$, there is a sequence of real numbers (initial conditions) ρ_n, with $\lim_{n\to\infty} \rho_n = 0$, and a corresponding sequence of real numbers t_n (times) such that the solution of (9.1)—call it $x_n(t)$—that satisfies $\|x_n(0)\| = \rho_n$ also satisfies $\|x_n(t_n)\| \geq \varepsilon$. Thus, not all solutions that start arbitrarily close to the trivial solution remain close to it for all future time. Note that it was important to take an entire sequence of initial conditions tending to zero. If the conclusion was satisfied for one, or a finite number of, initial conditions, there could be a smaller δ that would make the definition of stability "work" if solutions were this (δ) close. The infinite sequence of initial conditions guards against this possibility.

For example, the linear system

$$x' = \begin{bmatrix} 0 & 1 \\ 1 & 0 \end{bmatrix} x \tag{9.2}$$

has a fundamental matrix (actually e^{At}) of the form

$$\Phi = \begin{bmatrix} \cosh(t) & \sinh(t) \\ \sinh(t) & \cosh(t) \end{bmatrix}.$$

Using the theory we developed in Section 3, every solution of (9.2) can be written as $\Phi(t)c$, for some constant vector c. Choose $\rho_n = 1/n$, $n = 1, 2, \ldots$, and take c to be the vector

$$\begin{bmatrix} \dfrac{1}{2n} \\ \dfrac{1}{2n} \end{bmatrix}.$$

This corresponds to choosing the family of solutions

$$x_n(t) = \begin{bmatrix} \dfrac{1}{2n} \, e^t \\[2ex] \dfrac{1}{2n} \, e^t \end{bmatrix}.$$

Note that $\|x_n(0)\| = \rho_n = 1/n$. Take $\varepsilon = 1$ and $t_n = \ln(2n)$. After these elaborate preparations, we are ready to check the definition. Clearly, $\rho_n \to 0$ as $n \to \infty$. Yet $\|x_n(t_n)\| = 2 > 1 = \varepsilon$. Thus, a small change in initial condition at $t = 0$ produces a large ($> \varepsilon$) change at a future time, t_n. The formal definition of instability is satisfied (or, the definition of stability is violated).

A simple change in the system of equations (9.2) can make a dramatic change in the behavior of solutions. Consider the system

$$x' = \begin{bmatrix} 0 & 1 \\ -1 & 0 \end{bmatrix} x. \tag{9.3}$$

A fundamental matrix (e^{At}) is given by

$$= \begin{bmatrix} \cos(t) & \sin(t) \\ -\sin(t) & \cos(t) \end{bmatrix},$$

and every solution may be written as $\Phi(t)c$, or

$$x(t) = \begin{bmatrix} c_1 \cos(t) + c_2 \sin(t) \\ -c_1 \sin(t) + c_2 \cos(t) \end{bmatrix}$$

for appropriate constants c_1 and c_2. Let $\varepsilon > 0$ be given and choose $\delta = \varepsilon/2$. If $|c_1| + |c_2| < \delta$ (this is the norm of the vector $x(0)$—see Section 4 of this chapter), then $\|x(t)\| < 2(|c_1| + |c_2|) < 2\delta = \varepsilon$ for all $t > 0$ (in fact, for all t). Thus, the trivial solution of (9.3) is stable—solutions that begin sufficiently close to the trivial solution remain close to it in the future.

Although it is the simplest concept, it turns out that stability of the trivial solution is not the most important concept (in both mathematics and physics). The important concept is stronger and is called asymptotic stability. Its importance stems from the fact that it is preserved under slight changes ("perturbations") of A. If the trivial solution of (9.1) is asymptotically stable and if A is "changed" slightly, the trivial solution of the new (9.1) is also asymptotically stable. Since the entries of A often represent measured quantities, it is important that the stability property be retained under slight changes, corresponding, perhaps, to a measurement error. These ideas will be explored in detail in the next chapter for the two-dimensional case. Here we present the basic idea and a criterion for determining when the property holds. The trivial solution of (9.1) is said to be *asymptotically stable* if (a) it is stable, and (b) there is an $\varepsilon > 0$ such that if $\| x(0) \| < \varepsilon$, then $\lim_{t \to \infty} \| x(t) \| = 0$. Obviously, requirement (b) strengthens the conclusion. Neither system (9.2) nor (9.3) satisfies this condition. System (9.2), of course, satisfies neither (a) nor (b). However, the system

$$x' = \begin{bmatrix} -1 & 1 \\ 0 & -1 \end{bmatrix} x \tag{9.4}$$

has a fundamental matrix (again e^{At})

$$\Phi(t) = \begin{bmatrix} e^{-t} & te^{-t} \\ 0 & e^{-t} \end{bmatrix}.$$

Hence, every solution takes the form

$$x(t) = \begin{bmatrix} c_1 e^{-t} + c_2 te^{-t} \\ c_2 e^{-t} \end{bmatrix}, \tag{9.5}$$

and every solution satisfies $\lim_{t \to \infty} \| x(t) \| = 0$. This is strong enough to show that both (a) and (b) in the definition are satisfied, since the second component of the vector $x(t)$ is decreasing and the first component is eventually decreasing. The technical details are left as an exercise.

Fortunately, it is not necessary to solve each system to determine its stability. There is a simple theorem that provides a criterion for asymptotic stability for linear systems with constant coefficients.

THEOREM 9.1
   The trivial solution of (9.1) is asymptotically stable if and only if all of the eigenvalues of A have negative real parts.

When we say "negative real part" we intend that either the number is real and negative or it is complex and the real part is negative. A similar statement applies for "positive real part." There is a corresponding statement for instability.

THEOREM 9.2
   If one eigenvalue of A has a positive real part, then the trivial solution of (9.1) is unstable.

If the real part of the eigenvalues is nonpositive, the middle ground between asymptotic stability and instability is where the real part of at least one eigenvalue is zero. This case is more delicate and depends on the multiplicity of the eigenvalue with a zero real part. We will not give a detailed analysis, but note one simple result.

THEOREM 9.3

If the eigenvalues of A with zero real parts are simple and all other eigenvalues have negative real parts, then the trivial solution of (9.1) is stable.

In (9.3), both eigenvalues have zero real parts and the trivial solution is stable, but not asymptotically stable.

The key element in the proof of Theorem 9.1 is the establishment of the following basic lemma.

LEMMA 9.4

Let $\lambda_j = \xi_j + i\eta_j, j = 1, 2, \ldots, n$ be the eigenvalues of the matrix A (repetitions allowed). Let $\sigma > \max(\xi_j)$. Then there is a constant, $k > 0$, such that if $x(t)$ is a solution of (9.1), then

$$\|x(t)\| \le Ke^{\sigma t}.$$

The proof of this lemma is easy if A is a diagonal matrix and is not very difficult, using the properties of norms listed in Section 4, if A is similar to a diagonal matrix. For the general case we need to use the Putzer algorithm. We state the basic fact as another lemma.

LEMMA 9.5

Let A, σ, and λ_i be as in Lemma 9.4 and let $r_i(t)$ be the elements in the decomposition (7.1) of e^{At}. Then

$$|r_i(t)| \le c_i e^{\sigma t}$$

where c_i is a positive constant.

Proof. The proof is by induction. Let λ_i, $i = 1, 2, \ldots, n$ be given and let e^{At} be expressed by (7.1). Clearly, $|r_1(t)| = |e^{\lambda_1 t}| \le e^{\sigma t}$. Suppose that $|r_j(t)| \le c_j e^{\sigma t}, j = 1, 2, \ldots, i - 1$. Then, solving the scalar differential equation (or using the variation of constants formula, for the scalar—1×1 matrix—equation),

$$r_i' = \lambda_i r_i + r_{i-1}$$

yields (since $r_i(0) = 0$) that

$$r_i(t) = e^{\lambda_i t} \int_0^t e^{-\lambda_i s} r_{i-1}(s) \, ds.$$

It follows that

$$|r_i(t)| \leq \left| e^{\lambda_i t} \int_0^t e^{-\lambda_i s} r_{i-1}(s) \, ds \right|$$

$$\leq |e^{\lambda_i t}| \int_0^t |e^{-\lambda_i s}| |r_{i-1}(s)| \, ds$$

$$\leq c_{i-1} e^{\xi_i t} \int_0^t e^{(-\xi_i + \sigma)s} \, ds$$

since the real part of λ_i is ξ_i. After an integration,

$$|r_i(t)| \leq c_{i-1} e^{\xi_i t} \frac{e^{(\sigma - \xi_i)t} - 1}{\sigma - \xi}$$

$$\leq c_i e^{\sigma t}.$$

This completes the induction.

Proof of Lemma 9.4: Since the representation of the exponential e^{At} given by (7.1) has only a finite number of matrices, we can find a number, M, larger than the norm of each matrix. This number multiplied by the sum of the numbers c_i, $i = 1, 2, \ldots, n$, given by Lemma 9.5, provides the constant K in the statement of the lemma. To complete the proof, we use (7.1) and the inequalities for norms from Section 4 to find that

$$\|e^{At}\| \leq \sum_1^n |r_{i-1}(t)| \, \|P_i\|$$

$$< M e^{\sigma t} \sum_0^{n-1} c_i$$

$$\leq K e^{\sigma t}.$$

This completes the proof of Lemma 9.4.

With the aid of Lemma 9.4, the proof of Theorem 9.1 follows easily. Let $\sigma < 0$ be greater than the largest real part of any eigenvalue of A. This choice is possible because the largest eigenvalue has a negative real part. Since any solution $x(t)$ of (9.1) has the form $x(t) = e^{At} x(0)$, it follows from Lemma 9.4 that

$$\|x(t)\| \leq \|e^{At}\| \, \|x(0)\| \leq \|x(0)\| Ke^{\sigma t}, \qquad \sigma < 0.$$

Thus, $\lim_{t \to \infty} \|x(t)\| = 0$. This shows that **(b)** in the definition of asymptotic stability is satisfied for any solution. To get **(a)**, we have only to choose $\|x(0)\|$ small enough ($< \varepsilon/K$ in the definition of stability).

For the two-dimensional case, the fact that the largest eigenvalue has a negative real part can often be determined without actually computing the eigenvalues. For example, consider the system

$$x' = \begin{bmatrix} -2 & 4 \\ -3 & 1 \end{bmatrix} x. \tag{9.6}$$

The trace of the matrix gives the sum of the eigenvalues, while the determinant gives their product (quadratic formula). In (9.5) the trace is negative, so the sum of the eigenvalues is negative, while the product (the determinant) is positive. Thus, the eigenvalues have negative real parts. (They are, in fact, complex in this case.) For larger systems, criteria are known that guarantee that all of the eigenvalues of a matrix are negative or have negative real parts. Principal among these are the Routh-Hurwitz criteria, which the interested reader may find in more advanced textbooks.

It is important to note that things are somewhat more general than we have presented them to be. Stability of the trivial solution has been defined for (9.1). For other systems, and particularly for nonlinear ones, the stability of other types of solutions is important. However, for (9.1), if the matrix A is nonsingular, the only constant solution is the trivial one. In applied literature, these constant solutions are called *steady states* or *equilibrium solutions*. If A is singular, then there will be a "continuum" of solutions (a line, if A is two-dimensional). In this case, asymptotic stability of these constant solutions is not possible. Hence, the focus is on the zero solution. The definition was applied at the point in time $t = 0$. The definition could have been given for any other time t_0, but for equations of the form (9.1) a redefinition of initial time—moving t_0 to 0—is trivial. For other systems, the initial time t_0 may be crucial, and the definition is usually given with an arbitrary time t_0. The theorems are actually stronger than stated. For example, in the case of asymptotic stability the theorems are global in the sense that all solutions tend to zero as t tends to infinity, not just those that are initially close. Moreover, in view of Lemma 9.4, the rate of convergence to zero is exponential; that is, solutions tend to zero faster than does an exponential function. These properties are typical of linear systems but represent properties that are stronger than can be expected for other systems.

EXERCISES

1. Determine the stability of the trivial solution of $x' = 0$. Is it asymptotically stable?

2. Determine the asymptotic stability of the system $x' = Ax$ where A is

(a) $\begin{bmatrix} 1 & 2 \\ 3 & 4 \end{bmatrix}$
(b) $\begin{bmatrix} 1 & 0 \\ 0 & -1 \end{bmatrix}$

(c) $\begin{bmatrix} 1 & 1 & 0 \\ 0 & -2 & 1 \\ 0 & 0 & -1 \end{bmatrix}$
(d) $\begin{bmatrix} 1 & 1 & 1 \\ 1 & 0 & 1 \\ 0 & 0 & -2 \end{bmatrix}$

(e) $\begin{bmatrix} -1 & 0 & 0 \\ 0 & -2 & 0 \\ 1 & 0 & -1 \end{bmatrix}$
(f) $\begin{bmatrix} -1 & 0 & -1 \\ 0 & -2 & 0 \\ 1 & 0 & 0 \end{bmatrix}$

3. Let

$$A = \begin{bmatrix} a & b \\ c & d \end{bmatrix}.$$

Show that the sum of the eigenvalues is $a + d$ (called the trace of A) and the product is $ad - bc$ (the determinant of A). (*Hint:* Use the quadratic formula.)

4. Give a direct (i.e., without Lemma 9.5) proof of Lemma 9.4 when A is a diagonal matrix. (Use the form of e^{At} and the definition of norm from Section 4.)

5. Give a direct proof of Lemma 9.4 when $A = T^{-1}DT$ and D is a diagonal matrix.

6. Let $f(x)$ be a continuous function on the real line and c a real number such that $f(c) = 0$. Formulate a definition of stability for the constant solution $x(t) = c$ of $x' = f(x)$.

7. Repeat Exercise 6 for asymptotic stability.

8. Consider system (*) $x' = Ax + g(t)$, where A is an $n \times n$ constant matrix and $g(t)$ is a continuous n-dimensional vector. Use Theorem 8.3 and Lemma 9.4 to obtain the following estimate on the solution of (*) that satisfies $x(0) = x_0$,

$$\|x(t)\| \le K\|x_0\|e^{\sigma t} + Ke^{\sigma t} \int_0^t e^{-\sigma s}\|g(s)\|\, ds.$$

9. Suppose that A is a diagonal matrix and that the eigenvalues of A have negative real parts, except for one that has a zero real part. Use the form of e^{At} to show that all solutions of (9.1) are bounded. (By bounded we mean that there is a constant M, depending on the initial condition, such that $\|x(t)\| \le M$.)

10. Prove Theorem 9.3 in the special case that A is similar to a diagonal matrix.

11. Give a simple example to show that the statement in Exercise 10 is false if A has a double eigenvalue with zero real part and all others with negative real parts.

12. Consider (*) in Exercise 8 and suppose that the eigenvalues of A have negative real parts and that $\int_0^\infty \|g(t)\|\, dt$ exists. Show that all solutions of (*) are bounded. (*Hint:* Use the estimate in Exercise 8.)

13. Complete the formal details to show that the trivial solution of (9.4) is asymptotically stable. (*Hint:* Consider where te^{-t} has its maximum and use the triangle inequality.)

14. Redefine asymptotic stability using an arbitrary initial time $t = t_0$ instead of $t = 0$.

15. Show that the definition in Exercise 14 and that in the text are the same for the system (9.1). (*Hint:* Let $z(t) = x(t - t_0)$, find a differential equation for $z(t)$, and use Theorem 9.1.)

10. Periodic Coefficients

In the previous sections, the constant coefficient case played a prominent role. Many differential equations arise in a context that produces periodic rather than constant coefficients. The periodicity may reflect the variation of seasons in a population model or the rotation of the moon in a model of tidal activity. Moreover, certain stability questions for periodic solutions of nonlinear equations reduce to questions of the behavior of periodic linear systems. These equations have much in common with the constant coefficient case and the principal theorem lies close to the surface. In this section we give a brief description of the theory of such equations and state a simple theorem that has many applications.

Consider the linear system of differential equations

$$x' = A(t)x \tag{10.1}$$

where $A(t)$ is an $n \times n$ periodic matrix, that is, $A(t)$ satisfies

$$A(t + \omega) = A(t), \qquad t \in R. \tag{10.2}$$

The condition (10.2) is a wholesale way of saying that the matrix $A(t)$ consists of n^2 periodic functions with period ω. For convenience, we assume that ω is the smallest positive number for which (10.2) holds. We shall also assume that $A(t)$ is continuous on R. Since system (10.1) is linear, then the theory of Section 3 applies. In particular, all solutions of (10.2) can be obtained in the form

$$x(t) = \Phi(t)c$$

where Φ is a fundamental matrix and c is a constant vector. Knowledge of properties of a fundamental matrix then yields properties of solutions. A basic theorem in the theory of ordinary differential equations, known as *Floquet's theorem*, gives an important representation of a fundamental matrix when the coefficients are periodic.

THEOREM 10.1

Let (10.2) hold. If $\Phi(t)$ is a fundamental matrix for (10.1), then so is $\Psi(t) = \Phi(t + \omega)$, $t \in R$. Corresponding to the fundamental matrix $\Phi(t)$ there exists a periodic nonsingular matrix $P(t)$ with period ω and a constant matrix B such that

$$\Phi(t) = P(t)e^{Bt}. \tag{10.3}$$

The first statement is easy to prove. Let Ψ be defined as before and differentiate it to obtain

$$\Psi'(t) = \Phi'(t + \omega)$$
$$= A(t + \omega)\Phi(t + \omega)$$
$$= A(t)\Psi(t).$$

Moreover, $\det[\Psi(0)] = \det[\Phi(\omega)] \neq 0$. Thus Ψ is a fundamental matrix and the first statement is proved. While the proof of the remainder of the theorem is not difficult, it requires more linear algebra than we have developed. The reader is referred to standard texts for a complete proof. The exercises give an indication of the proof for some special cases.

The importance of the theorem is that it factors a fundamental matrix into a periodic part and a "stability" part. The determination of a fundamental matrix over an interval $[0, \omega]$ determines it for all values of t. Moreover, from previous material we know something of the structure of matrices of the form e^{Bt} and we wish to apply this to obtain information about solutions of system (10.1). As an example, consider the question of whether there are any periodic solutions of system (10.1) of period ω. If $y(t)$ is a nontrivial solution of (10.1), then $y(t) = \Phi(t)c$ for some nonzero vector c. That $y(t) = y(t + \omega)$ takes the form

$$\Phi(t)c = \Phi(t + \omega)c.$$

An application of Theorem 10.1 changes this to

$$P(t)e^{Bt}c = P(t + \omega)e^{B(t+\omega)}c$$

or

$$P(t)[e^{B(t+\omega)} - e^{Bt}]c = 0.$$

Using Exercise 5 of Section 4, we have that

$$P(t)e^{Bt}[e^{B\omega} - I]c = 0.$$

Since $P(t)$ and e^{Bt} are nonsingular, it follows that the determinant of the matrix in square brackets must be singular or c must be the zero vector. Since we are assuming that $y(t)$ was a nontrivial solution, it must be the case that

$$\det [e^{B\omega} - I] = 0.$$

This in turn says that 1 must be an eigenvalue of the matrix $e^{B\omega}$. The steps above are reversible, so this condition is both necessary and sufficient. The eigenvalues of $e^{B\omega}$ are called the *Floquet multipliers*. The preceding discussion may be sumarized by the statement that (10.1) *has a periodic solution of period ω if and only if a Floquet multiplier is equal to* 1. The question of whether all solutions of (10.1) are periodic with period ω is more difficult. One such case is discussed in the exercises.

In general, determining the Floquet multipliers is an impossible task. The following observation is sometimes useful. Let $\Phi(t)$ be a fundamental matrix that satisfies $\Phi(0) = I$. Then, from Theorem 10.1, it follows that

$$\begin{aligned} \Phi(t + \omega) &= P(t + \omega)e^{B(t+\omega)} \\ &= P(t)e^{Bt}e^{B\omega} \\ &= \Phi(t)e^{B\omega}. \end{aligned}$$

Setting $t = 0$ yields that

$$\Phi(\omega) = e^{B\omega}, \tag{10.4}$$

or that finding the Floquet multipliers is the same as finding the eigenvalues of a (particular) fundamental matrix evaluated at the period.

System (10.1) has much in common with the constant coefficient case. The following lemma shows the connection in a very explicit way.

LEMMA 10.2

There exists a (periodic) change of variables that takes solutions of system (10.1) onto solutions of the constant coefficient system

$$y' = By \tag{10.5}$$

where B is given in Theorem 10.1.

Proof. Fix Φ, a fundamental matrix for (10.1), with $\Phi(0) = I$. Let $P(t)$ and B be given by Theorem 10.1 and let $P^{-1}(t)$ be the inverse of $P(t)$. Define a new dependent variable $y(t)$ by $y(t) = P^{-1}(t)x(t)$. Since $P(t) = \Phi(t)e^{-Bt}$, it follows that

$$P'(t) = \Phi'(t)e^{-Bt} + \Phi(t)[e^{-Bt}]'$$
$$= A(t)\Phi(t)e^{-Bt} - \Phi(t)e^{-Bt}B$$
$$= A(t)P(t) - P(t)B$$

where Theorem 4.3 has been used in the second step. Now

$$x'(t) = [P(t)y(t)]'$$
$$= P'(t)y(t) + P(t)y'(t)$$
$$= [A(t)P(t) - P(t)B]y(t) + P(t)y'(t).$$

Using the differential equation (10.1) and the transformation yields that

$$A(t)P(t)y(t) = [A(t)P(t) - P(t)B]y(t) + P(t)y'(t).$$

This expression simplifies to

$$y'(t) = By(t), \tag{10.6}$$

since P is invertible. This establishes the lemma.

The significance of this result is that the behavior of solutions of (10.1) can be determined by the behavior of the constant coefficient case (10.6). The behavior of solutions of (10.6) is determined by the eigenvalues of the matrix B. The practical difficulty of this approach, of course, is that we do not really have the matrix $P(t)$. Its existence is given by Theorem 10.1, but this theorem is not constructive. It is useful, however, for theoretical purposes. The eigenvalues of B are called the *Floquet exponents*. The concepts of multipliers and exponents are connected, as shown in the following lemma (whose proof is omitted).

LEMMA 10.3

If λ_i is a **Floquet multiplier for the system (10.1), then there is a corresponding Floquet exponent** ρ_i **such that**

$$\lambda_i = e^{\omega\rho_i}.$$

If we adopt the same definition of stability of the trivial solution for (10.1) that we used in the preceding section (this is possible, since A is periodic), then the following holds.

THEOREM 10.4

If the Floquet multipliers λ_i satisfy $|\lambda_i| < 1$, $i = 1, \ldots, n$, the trivial solution of (10.1) is asymptotically stable.

Proof. We show first that it is sufficient to establish that the trivial solution of (10.6) is asymptotically stable. Since the entries of $P(t)$ are continuous, $\|P(t)\|$ is a continuous function on $[0, \omega]$, and hence is bounded there. Since it is periodic of period ω, it is bounded for all t. Call the bound M. Thus,

$$\|x(t)\| \leq \|P(t)\|\,\|y(t)\|$$

$$\leq M\|y(t)\|.$$

Hence, if $\|y(t)\|$ tends to zero as t tends to infinity, so does $\|x(t)\|$. Moreover, if $\|y(t)\|$ remains small, so does $\|x(t)\|$ (smallness is adjusted by the constant M). To make this last statement precise, we carry out the technical verification of stability. Let $\varepsilon > 0$ be given. Suppose that the trivial solution of (10.6) is (asymptotically) stable. Then, for every $\varepsilon > 0$ there is a $\delta > 0$ such that if $\|y(0)\| < \delta$, then $\|y(t)\| < \varepsilon$ for all $t > 0$. Apply this definition to $y(t)$ with $\varepsilon_1 = \varepsilon/M$. Use the resulting δ for the δ needed in the definition of stability for $x(t)$. Then

$$\|x(t)\| \leq M\|y(t)\|$$

$$< M\varepsilon_1$$

$$< M\frac{\varepsilon}{M} = \varepsilon.$$

Thus the trivial solution of (10.1) is stable. The asymptotic part will follow directly, as noted above.

Equation (10.6) will be asymptotically stable if and only if the eigenvalues of B have negative real parts. Denote the eigenvalues of B by ρ_i. From Lemma 10.3, it follows that

$$e^{\omega(\text{Real part } \rho_i)} = |\lambda_i| < 1.$$

Hence, the real parts of the Floquet exponents must be negative. Thus, the trivial solution of (10.6) is asymptotically stable by Theorem 9.1, and hence so is the trivial solution of (10.1).

EXERCISES

In working all of the following problems, assume that (10.2) holds.

1. Show that there is a matrix C such that $\Phi(t)C = \Phi(t + \omega)$, where Φ is a fundamental matrix for the system (10.1)

2. If C is an $n \times n$ nonsingular, diagonal matrix, show that there is a matrix B such that $e^B = C$.

3. If C is a nonsingular $n \times n$ matrix that is similar to a diagonal matrix, show that there is a matrix B such that $e^B = C$. (*Hint:* Use equation (4.4).)

4. Suppose that if C is nonsingular, there exists a matrix B such that $e^B = C$. The following is the crucial step in the proof of Theorem 10.1.
      Let Φ be a fundamental matrix for (10.1), let B be such that $C = e^{B\omega}$, and let $P(t) = \Phi(t)e^{-Bt}$. Show that $P(t)$ is periodic with period ω.

5. Deduce the conclusion of Theorem 10.1 from Exercise 4.

6. Suppose that $A(t)$ is odd—that is, $A(t) = -A(-t)$. Show that a fundamental matrix Φ of (10.1) satisfies $\Phi(t) = \Phi(-t)$. (*Hint:* Show that $\Phi(t)$ and $\Phi(-t)$ satisfy the same differential equation and initial condition.)

7. Let A and Φ be as in Exercise 6. Show that $\Phi(t - \omega) = \Phi(t)$. (*Hint:* Use Exercises 1 and 6.)

8. Use Exercises 6 and 7 to show that if $A(t)$ is odd, then all solutions of (10.1) are periodic with period ω.

9. Consider the special case

$$ x' = \begin{vmatrix} 0 & 1 \\ p(t) & 0 \end{vmatrix} x $$

where $p(t)$ has period ω. Let $\Phi(t)$ be a fundamental matrix such that $\Phi(0) = I$. Show that $\det[\Phi(t)] = 1$ for every t. Use this fact to show that the Floquet multipliers are roots of $\lambda^2 - \mu\lambda + 1 = 0$ where μ is the trace of $\Phi(\omega)$.

10. If $\mu = 2$, show that the differential equation in Exercise 9 has a solution of period ω.

11. If $-2 < \mu < 2$, show that all solutions of the equation in Exercise 9 are bounded. Can you say more if, in addition, $\mu \neq 0$?

11.   Scalar Equations

Section 10 concluded the basic development of the theory of systems of linear differential equations. In the introduction to this chapter the equivalence of the scalar equation

$$y'' + a(t)y' + b(t)y = e(t) \tag{11.1}$$

and the system

$$z' = \begin{bmatrix} 0 & 1 \\ -b(t) & -a(t) \end{bmatrix} z + \begin{bmatrix} 0 \\ e(t) \end{bmatrix} \tag{11.2}$$

was demonstrated. It was shown that if $y(t)$ is a solution of (11.1), then the vector

$$z(t) = \begin{bmatrix} y(t) \\ y'(t) \end{bmatrix}$$

is a solution of (11.2). Conversely, if

$$z(t) = \begin{bmatrix} z_1(t) \\ z_2(t) \end{bmatrix}$$

is a solution of the system (11.2), then $z_1(t)$ is a solution of the scalar equation (11.1). A similar result holds for the nth-order scalar equation

$$y^{(n)} + a_1(t)y^{(n-1)} + \cdots + a_n(t)y = e(t) \tag{11.3}$$

and the n-dimensional system

$$z' = \begin{bmatrix} 0 & 1 & 0 & 0 & \cdots & 0 \\ 0 & 0 & 1 & 0 & \cdots & 0 \\ \vdots & & & & & \vdots \\ & & & & & 1 \\ -a_n(t) & -a_{n-1}(t) & & \cdots & & -a_1(t) \end{bmatrix} z + \begin{bmatrix} 0 \\ 0 \\ \vdots \\ e(t) \end{bmatrix} . \tag{11.4}$$

The purpose of this section is to briefly relate elements of theory just developed for systems to the more familiar case of scalar equations studied in a previous course. It should be emphasized that changing a scalar equation into a system to use this theory is generally not efficient. The theory for scalar linear differential equations is best applied directly to the nth-order scalar equation. The changes of form considered here are for illustrative purposes only. To illustrate the connection, we use second-order equations with constant coefficients.

To find solutions of the scalar equation

$$y'' + ay' + by = 0, \tag{11.5}$$

with a and b constant, we first find the roots of the polynomial, usually called the characteristic polynomial,

$$\lambda^2 + a\lambda + b = 0. \tag{11.6}$$

The corresponding system, in the form $z' = Az$, is

$$z' = \begin{bmatrix} 0 & 1 \\ -b & -a \end{bmatrix} z. \tag{11.7}$$

The eigenvalues of the coefficient matrix are given by

$$\det \begin{bmatrix} -\lambda & 1 \\ -b & -a - \lambda \end{bmatrix} = 0$$

or by

$$(-\lambda)(-a - \lambda) + b = 0.$$

This simplifies to

$$\lambda^2 + a\lambda + b = 0, \tag{11.8}$$

which is the same as (11.6). Hence, the eigenvalues of the coefficient matrix A of the corresponding system and the roots of the characteristic polynomial for scalar equations are the same. An eigenvector corresponding to the eigenvalue λ of the form

$$\begin{bmatrix} c_1 \\ c_2 \end{bmatrix}$$

has entries that satisfy

$$\begin{aligned} -\lambda c_1 + c_2 &= 0 \\ -b c_1 - (a + \lambda) c_2 &= 0. \end{aligned} \tag{11.9}$$

Hence, $c_2 = \lambda c_1$ and the eigenvector may be taken to be of the form

$$\begin{bmatrix} 1 \\ \lambda \end{bmatrix}.$$

Suppose first that there are two distinct roots of (11.8), λ_1 and λ_2. Then, a fundamental matrix for (11.7) takes the form

$$\Phi(t) = \begin{bmatrix} e^{\lambda_1 t} & e^{\lambda_2 t} \\ \lambda_1 e^{\lambda_2 t} & \lambda_2 e^{\lambda_2 t} \end{bmatrix}.$$

Equation (11.5) has two linearly independent solutions, $e^{\lambda_1 t}$ and $e^{\lambda_2 t}$. If $y_1(t)$ and $y_2(t)$ are two solutions of (11.5), recall that the Wronskian is defined to be

$$\det \begin{bmatrix} y_1(t) & y_2(t) \\ y_1'(t) & y_2'(t) \end{bmatrix}.$$

In this case, the Wronskian takes the form

$$W = \lambda_2 e^{\lambda_2 t} e^{\lambda_1 t} - \lambda_1 e^{\lambda_1 t} e^{\lambda_2 t}.$$

Thus, the Wronskian of two appropriately selected, linearly independent solutions of (11.5) corresponds to the determinant of a fundamental matrix constructed by using the solutions of the scalar equation as entries in the first row.

In the case of a repeated eigenvalue, ($\lambda_1 = \lambda_2 = \lambda$), the situation is somewhat more complicated. First of all, using the quadratic formula we see that the eigenvalues are equal if and only if $a^2 - 4b = 0$. This, in turn, implies that $\lambda = -a/2$ and $\lambda^2 = b$. Two linearly independent solutions of the scalar equation (11.5) are given by $y_1(t) = e^{\lambda t}$ and $y_2(t) = te^{\lambda t}$. There will always be just one (linearly independent) eigenvector since, as a consequence of (11.8), we must always have $c_2 = \lambda c_1$. To find a solution in this case, we construct e^{At}. Because of the form of the matrix, this is not difficult. Following the Putzer algorithm (Theorem 7.1), we note that

$$P_0 = I$$

$$P_1 = (A - \lambda I) = \begin{bmatrix} -\lambda & 1 \\ -b & -a - \lambda \end{bmatrix}$$

$$r_1(t) = e^{\lambda t}$$

$$r_2(t) = te^{\lambda t}.$$

Thus e^{At} takes the form

$$\begin{bmatrix} e^{\lambda t} - \lambda te^{\lambda t} & te^{\lambda t} \\ -bte^{\lambda t} & e^{\lambda t} + te^{\lambda t}(-a - \lambda) \end{bmatrix}$$

or, using the form of λ,

$$e^{At} = \begin{bmatrix} e^{\lambda t} - \lambda t e^{\lambda t} & t e^{\lambda t} \\ -\lambda^2 t e^{\lambda t} & e^{\lambda t} + \lambda t e^{\lambda t} \end{bmatrix}. \tag{11.10}$$

The determinant of this fundamental matrix is not the Wronskian of the two solutions of (11.5) chosen above, but this is a matter of the choice of those solutions. If we had chosen $y_1(t) = e^{\lambda t} - \lambda t e^{\lambda t}$ and $y_2(t) = t e^{\lambda t}$ as the linearly independent solutions of (1.5), then the Wronskian and the determinant of the fundamental matrix (11.10) would have been the same. Alternatively, we could choose as a fundamental matrix $e^{At}C$, where

$$C = \begin{bmatrix} 1 & 0 \\ \lambda & 1 \end{bmatrix}.$$

Then $e^{At}C$ has $e^{\lambda t}$ and $t e^{\lambda t}$ as elements of the first row. Obviously, if we choose the linearly independent solutions $y_1(t)$, $y_2(t)$, of the scalar equation to be "normalized" to $y_1(0) = 1$, $y_1'(0) = 0$, $y_2(0) = 0$, $y_2'(0) = 1$, and choose e^{At} as the fundamental matrix for the system, then the first row of the fundamental matrix will be the two linearly independent solutions of the scalar equation.

In a similar way, we could carry out the correspondence between the nth-order scalar equivalence and the n-dimensional fundamental matrix. The linear algebra becomes more complicated, but the basic idea is the same. The linearly independent solutions of the scalar equation can form the first row of the fundamental matrix of the corresponding system.

The stability theory of Section 9 carries over to the scalar equation as a consequence of the equivalence above. We have that if the real parts of the roots of (11.6) are negative, then all solutions of (11.5) tend to zero as t tends to infinity. We leave the formal proof of this fact as an exercise.

The theory for periodic coefficients can be similarly translated into a theory for scalar equations with periodic coefficients. The equation

$$y'' + a(t)y = 0,$$

with $a(t)$ periodic, is particularly important in applied mathematics and has a very detailed theory. The exercises explore the first steps of the Floquet theorem for this special case.

EXERCISES

1. Show the equivalence of the solutions of (11.3) and (11.4).

2. Change $y''' + 3y'' + y' + y = 0$ to a system of the form $z' = Az$. Solve the scalar equation and find e^{At}.

3. Use the system form of $y'' + y = 0$ to show that the Wronskian of any two solutions is constant.

4. Show that if the real parts of all solutions of (11.6) are negative, then $\lim_{t\to\infty} y(t) = 0$ for every solution of (11.5).

5. Extend Exercise 4 to the nth-order linear equation with constant coefficients.

6. Formulate a definition of stability for (11.5) based on the system form of the equation.

7. Interpret $y(0)$ as the initial position and $y'(0)$ as the initial velocity and repeat the definition in Exercise 6 in these terms (in English, without symbols).

Consider the equation

$$y'' + a(t)y = 0 \qquad\qquad (*)$$

with $a(t)$ periodic with period ω. The terms "Floquet multiplier" and "Floquet exponent" in the following exercises refer to the system form of this equation.

8. Show that the product of the Floquet multipliers of (*) is equal to 1. (*Hint:* Note that the trace of the corresponding system is zero.)

9. Show that if $y(t)$ is a solution of (*), then so is $y(t + \omega)$. Let $y_1(t)$ and $y_2(t)$ be two linearly independent solutions of (*) satisfying normalized initial conditions $y_1(0) = 1$, $y_1'(0) = 0$, $y_2(0) = 0$, $y_2'(0) = 1$. Show that the multipliers satisfy

$$r^2 - [y_1(\omega) + y_2'(\omega)]r + 1 = 0.$$

10. Let y_1 and y_2 be as in Exercise 9. Show that (*) has a solution that satisfies

$$y(t + \omega) = ry(t) \qquad\qquad (**)$$

if and only if r is a Floquet multiplier. (*Hint:* $y_1(t + \omega)$, $y_2(t + \omega)$, and $y(t)$ can be expressed as a linear combination of $y_1(t)$ and $y_2(t)$. Use (**) and Exercise 9.) Equation (**) suggests why the term "multiplier" is used.

11. Clearly, (*) has a solution of period T if and only if $r = 1$. Show that (*) has a solution of least period $2T$ if and only if $r = -1$.

12. Let $y_1(t)$ and $y_2(t)$ be two (not necessarily distinct) solutions of (*). Show that $z(t) = y_1(t)y_2(t)$ is a solution of

$$z''' + 4q(t)z' + 2q'(t)z = 0. \qquad\qquad (***)$$

Conclude that y_1^2, y_2^2, and $y_1 y_2$ are solutions of (***).

12. An Application: Coupled Oscillators

A mass, suspended from a rod, forms a simple pendulum (Figure 12.1). As long as the oscillations are small, the motion is approximated by the solutions

Figure 12.1 The simple pendulum.

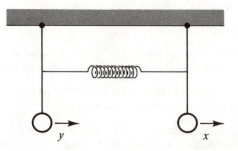

Figure 12.2 Two pendulums coupled with a spring.

of the second-order linear differential equation

$$m\theta'' + m\omega_0^2\theta = 0,$$

with initial condition $\theta(0) = \theta_0$, $\theta'(0) = v_0$. The initial conditions $\theta(0)$, $\theta'(0)$ correspond to the initial angle of the pendulum, that is, its angle at $t = 0$, and the initial (angular) velocity of the pendulum. The constant ω_0^2 is determined by the length of the pendulum, L, and the gravitational constant, g, as $\omega_0^2 = g/L$.

Consider now two pendulums of equal length and equal mass coupled by a spring (Figure 12.2). If the spring was not present, the motion of each pendulum would be governed by equations of the form above. Let $x(t)$ and $y(t)$ denote the respective angles at time t. The action of the spring is to introduce a force proportional to the difference of the displacements, a force $\hat{k}(x - y)$. The motion then is described by solutions of

$$mx'' + m\omega_0^2 x = -\hat{k}(x - y)$$
$$my'' + m\omega_0^2 y = -\hat{k}(y - x),$$

(12.1)

which we write in system form

$$x_1' = x_2$$

$$x_2' = -\omega_0^2 x_1 - kx_1 + ky_1$$

$$y_1' = y_2$$

$$y_2' = -\omega_0^2 y_1 - ky_1 + kx_1,$$

where $k = \hat{k}/m$. This is a system of the form

$$z' = Az \tag{12.2}$$

where

$$A = \begin{bmatrix} 0 & 1 & 0 & 0 \\ -\omega_0^2 - k & 0 & k & 0 \\ 0 & 0 & 0 & 1 \\ k & 0 & -\omega_0^2 - k & 0 \end{bmatrix}$$

and

$$z = \begin{bmatrix} x_1 \\ x_2 \\ y_1 \\ y_2 \end{bmatrix}.$$

We apply the techniques of this chapter to find a fundamental matrix and thereby to describe all the solutions of (12.2). First, the determinant

$$\det \begin{bmatrix} -\lambda & 1 & 0 & 0 \\ -(\omega_0^2 + k) & -\lambda & k & 0 \\ 0 & 0 & -\lambda & 1 \\ k & 0 & -(\omega_0^2 + k) & -\lambda \end{bmatrix}$$

can be expanded along the first row to give

$$-\lambda \begin{bmatrix} -\lambda & k & 0 \\ 0 & -\lambda & 1 \\ 0 & -(\omega_0^2 + k) & -\lambda \end{bmatrix} - \begin{bmatrix} -(\omega_0^2 + k) & k & 0 \\ 0 & -\lambda & 1 \\ k & -(\omega_0^2 + k) & -\lambda \end{bmatrix}$$

or

$$\lambda^2(\lambda^2 + \omega_0^2 + k) + (\omega_0^2 + k)(\lambda^2 + \omega_0^2 + k) - k^2.$$

The characteristic roots are the solutions of

$$(\lambda^2 + \omega_0^2 + k)^2 - k^2 = 0$$

or

$$\lambda = \pm i\omega_0$$
$$\lambda = \pm i\sqrt{\omega_0^2 + 2k}.$$

The eigenvector corresponding to $i\omega_0$ is a solution of

$$\begin{bmatrix} -i\omega_0 & 1 & 0 & 0 \\ -\alpha & -i\omega_0 & k & 0 \\ 0 & 0 & -i\omega_0 & 1 \\ k & 0 & -\alpha & -i\omega_0 \end{bmatrix} \begin{bmatrix} c_1 \\ c_2 \\ c_3 \\ c_4 \end{bmatrix} = 0$$

where $\alpha = \omega_0^2 + k$. If we set $c_1 = 1$ (arbitrarily), then

$$c_2 = i\omega_0.$$

Then c_3 can be found from

$$-\alpha c_1 - i\omega_0 c_2 + k c_3 = 0,$$

given by

$$c_3 = \frac{\alpha - \omega_0^2}{k} = 1.$$

Finally,

$$-i\omega_0 c_3 + c_4 = 0$$

or

$$c_4 = i\omega_0.$$

Thus, one solution is given by

$$e^{i\omega_0 t} \begin{pmatrix} 1 \\ i\omega_0 \\ 1 \\ i\omega_0 \end{pmatrix}.$$

Taking the real and imaginary parts and using Euler's formula, two solutions are given by the vectors

$$
\begin{pmatrix} \cos(\omega_0 t) \\ -\omega_0 \sin(\omega_0 t) \\ \cos(\omega_0 t) \\ -\omega_0 \sin(\omega_0 t) \end{pmatrix} \quad \text{and} \quad \begin{pmatrix} \sin(\omega_0 t) \\ \omega_0 \sin(\omega_0 t) \\ \sin(\omega_0 t) \\ \omega_0 \cos(\omega_0 t) \end{pmatrix}.
$$

Let $\omega_1 = \sqrt{\omega_0^2 + 2k}$. Then a second eigenvector can be found in the same way as a solution of

$$
\begin{pmatrix} -i\omega_1 & 1 & 0 & 0 \\ -\alpha & -i\omega_1 & k & 0 \\ 0 & 0 & -i\omega_1 & 1 \\ -k & 0 & -\alpha & -i\omega_1 \end{pmatrix} \begin{pmatrix} c_1 \\ c_2 \\ c_3 \\ c_4 \end{pmatrix} = 0.
$$

Proceeding as above, set $c_1 = 1$ arbitrarily and find that $c_2 = i\omega_1$. Then c_3 is a solution of

$$
-\alpha + \omega_1^2 + kc_3 = 0
$$

or

$$
c_3 = \frac{\alpha - \omega_1^2}{k} = \frac{\omega_0^2 + k - \omega_1^2}{k}
$$

$$
= \frac{\omega_0^2 + k - \omega_0^2 - 2k}{k} = -1.
$$

It follows from

$$
-i\omega_1 c_3 + c_4 = 0
$$

that

$$
c_4 = i\omega_1 c_3 = -i\omega_1.
$$

Hence, two solutions are given by the real and imaginary parts of

$$
e^{i\omega_1 t} \begin{pmatrix} 1 \\ i\omega_1 \\ -1 \\ -i\omega_1 \end{pmatrix},
$$

which are

$$
\begin{pmatrix}
\cos(\omega_1 t) \\
-\omega_1 \sin(\omega_1 t) \\
-\cos(\omega_1 t) \\
\omega_1 \sin(\omega_1 t)
\end{pmatrix}
\quad \text{and} \quad
\begin{pmatrix}
\sin(\omega_1 t) \\
\omega_1 \cos(\omega_1 t) \\
-\sin(\omega_1 t) \\
-\omega_1 \cos(\omega_1 t)
\end{pmatrix}.
$$

Using these four solutions, we construct a solution of the corresponding matrix equation to be

$$
\Phi(t) =
\begin{bmatrix}
\cos(\omega_0 t) & \sin(\omega_0 t) & \cos(\omega_1 t) & \sin(\omega_1 t) \\
-\omega_0 \sin(\omega_0 t) & \omega_0 \cos(\omega_0 t) & -\omega_1 \sin(\omega_1 t) & \omega_1 \cos(\omega_1 t) \\
\cos(\omega_0 t) & \sin(\omega_0 t) & -\cos(\omega_1 t) & -\sin(\omega_1 t) \\
-\omega_0 \sin(\omega_0 t) & \omega_0 \cos(\omega_0 t) & \omega_1 \sin(\omega_1 t) & -\omega_1 \cos(\omega_1 t)
\end{bmatrix}.
$$

$$(12.3)$$

Since

$$
\Phi(0) =
\begin{bmatrix}
1 & 0 & 1 & 0 \\
0 & \omega_0 & 0 & \omega_1 \\
1 & 0 & -1 & 0 \\
0 & \omega_0 & 0 & -\omega_1
\end{bmatrix},
$$

$\det[\Phi(0)] = 4\omega_0\omega_1 \neq 0$, so Φ is a *fundamental matrix* and all solutions are of the form

$$
\Phi(t)c \tag{12.4}
$$

for an appropriate vector c.

To clarify the nature of the resulting motion, let us consider some special initial conditions. First, suppose that the initial condition in (12.2) is taken as

$$
z(0) =
\begin{pmatrix}
1 \\
0 \\
1 \\
0
\end{pmatrix}.
$$

This corresponds to an initial condition in (12.1) of $x_1(0) = 1$, $x_1'(0) = 0$, $y_1(0) = 1$, $y_1'(0) = 0$, equal displacements of the pendulums, and no initial velocities. Then, to find c of (12.4), we solve

$$
\begin{bmatrix} 1 & 0 & 1 & 0 \\ 0 & \omega_0 & 0 & \omega_1 \\ 1 & 0 & -1 & 0 \\ 0 & \omega_0 & 0 & -\omega_1 \end{bmatrix} \begin{bmatrix} c_1 \\ c_2 \\ c_3 \\ c_4 \end{bmatrix} = \begin{bmatrix} 1 \\ 0 \\ 1 \\ 0 \end{bmatrix}
$$

to find that

$$
c = \begin{bmatrix} 1 \\ 0 \\ 0 \\ 0 \end{bmatrix}.
$$

The solution corresponding to this initial condition is $z(t) = \Phi(t)c$, which is just the first column of (12.3), or

$$
z(t) = \begin{bmatrix} \cos(\omega_0 t) \\ -\omega_0 \sin(\omega_0 t) \\ \cos(\omega_0 t) \\ -\omega_0 \sin(\omega_0 t) \end{bmatrix}.
$$

In terms of the original equations (12.1), this is

$$
x(t) = \cos(\omega_0 t)
$$
$$
y(t) = \cos(\omega_0 t).
$$

This solution is periodic with period $2\pi/\omega_0$ and corresponds to the case that the motion of the pendulums is such that *the spring is never stretched.*

Now consider the initial condition

$$
z(0) = \begin{pmatrix} 1 \\ 0 \\ -1 \\ 0 \end{pmatrix}.
$$

Working out $\Phi(0)c = z(0)$, we find that

$$
c = \begin{bmatrix} 0 \\ 0 \\ -1 \\ 0 \end{bmatrix},
$$

and that

$$z(t) = - \begin{bmatrix} \cos(\omega_1 t) \\ \omega_1 \sin(\omega_1 t) \\ -\cos(\omega_1 t) \\ \omega_1 \sin(\omega_1 t) \end{bmatrix}.$$

In terms of the original variables, $x(t) = -\cos(\omega_1 t)$ and $y(t) = \cos(\omega_1 t)$. The motion of the two pendulums is exactly opposite and the resulting motion is periodic with period $2\pi/\omega_1$. Since $\omega_1^2 = \omega_0^2 + 2k$, $\omega_1 > \omega_0$, so the period of oscillations is decreased—*the action of the spring is such that it makes the pendulums oscillate faster.*

Consider now an initial condition

$$z(0) = \begin{pmatrix} 1 \\ 0 \\ 0 \\ 0 \end{pmatrix},$$

which corresponds to starting with one pendulum at the equilibrium position and displacing the other. We must solve $\Phi(0)c = z(0)$, or

$$\begin{bmatrix} 1 & 0 & 1 & 0 \\ 0 & \omega_0 & 0 & \omega_1 \\ 1 & 0 & -1 & 0 \\ 0 & \omega_0 & 0 & -\omega_1 \end{bmatrix} \begin{bmatrix} c_1 \\ c_2 \\ c_3 \\ c_4 \end{bmatrix} = \begin{bmatrix} 1 \\ 0 \\ 0 \\ 0 \end{bmatrix}.$$

The solution is

$$c = \begin{bmatrix} \tfrac{1}{2} \\ 0 \\ \tfrac{1}{2} \\ 0 \end{bmatrix},$$

corresponding to a solution of the differential equation of

$$z(t) = \frac{1}{2} \begin{pmatrix} \cos(\omega_0 t) + \cos(\omega_1 t) \\ -\omega_0 \sin(\omega_0 t) + \omega_0 \sin(\omega_1 t) \\ \cos(\omega_0 t) - \cos(\omega_1 t) \\ -\omega_0 \sin(\omega_0 t) - \omega_1 \sin(\omega_1 t) \end{pmatrix}.$$

In terms of the original variables, this is

$$x(t) = \frac{\cos(\omega_0 t) + \cos(\omega_1 t)}{2}$$

$$y(t) = \frac{\cos(\omega_0 t) - \cos(\omega_1 t)}{2}.$$

The solution is a combination of two periodic functions with frequencies ω_0 and ω_1 and will *not* be periodic unless k is such that ω_1 is a rational multiple of ω_0. Such a solution falls in the class of *almost periodic* functions, a topic that we cannot pursue here.

In summary, the solutions may be periodic with frequencies ω_0 or $\omega_1 > \omega_0$, or may be a combination of two periodic functions, one with period ω_0 and one with period ω_1. We can think of ω_0 as the natural frequency, without the spring, and of ω_1 as the frequency introduced by the spring. Although this combination of functions will be oscillatory in the sense of having arbitrarily large positive maxima and negative minima, the motion need not be periodic.

EXERCISES

1. Let $\omega_1 > 0$ and $\omega_2 > 0$. Show that $f(t) = \sin(\omega_1 t) + \sin(\omega_2 t)$ is periodic if $a\omega_1 + b\omega_2 = 0$, a, b integers. Find the period.

2. Let $f(t)$ be as in Exercise 1. Suppose that $a\omega_1 + b\omega_2 \neq 0$ for any integers a and b $(a, b) \neq (0, 0)$. Show that $f(t)$ is not periodic. (*Hint:* Suppose the contrary, then differentiate, evaluate at $t = 0$, and argue to a contradiction.)

3. Suppose the masses of the two pendulums is Figure 12.2 are different. Modify the equations and find a fundamental matrix.

4. Suppose the lengths of the two pendulums are different. Modify the equations and find a fundamental matrix (complicated).

REFERENCES

The theory of linear systems of ordinary differential equations appears in all of the standard graduate textbooks. Two examples are

Coddington, E. A., and N. Levinson. *Theory of Ordinary Differential Equations.* New York: McGraw-Hill, 1955.

Hale, J. K. *Ordinary Differential Equations.* New York: Wiley-Interscience, 1969.

A more intermediate-level book, emphasizing systems and stability, is

Brauer, F., and J. A. Nohel. *Qualitative Theory of Ordinary Differential Equations.* Menlo Park, CA: The Benjamin/Cummings Publishing Company, 1969.

The Putzer algorithm appears in

Putzer, E. J. "Avoiding the Jordan Conical Form in the Discussion of Linear Systems with Constant Coefficients." *Amer. Math. Monthly* 73 (1966): 2–7.

Stability theory appears in very elegant form in

Coppel, W. A. *Stability and Asymptotic Behavior of Differential Equations.* Lexington, MA: D. C. Heath & Company, 1965.

A very complete study of equations with periodic coefficients is given in

Eastham, M. S. P. *The Spectral Theory of Periodic Differential Equations.* Edinburgh: Scottish Academic Press, 1973.

A classic reference for the scalar case is

Magnus, W., and S. Winkler. *Hill's Equation.* New York: Wiley-Interscience, 1966.

The example of the coupled oscillators appears in

Feynman, R. P., R. B. Leighton, and M. Sands. *The Feynman Lectures on Physics.* Vol. 1. Reading, MA: Addison-Wesley, 1965.

2

Two-Dimensional Autonomous Systems

1. Introduction

The preceding chapter dealt with linear differential equations. Although this theory is very important, the equations that arise in most physical and biological systems are inherently nonlinear. However, the qualitative behavior of solutions of nonlinear differential equations near "stationary" solutions (to be defined later) can, on some occasions, be determined from the behavior of linear approximations. Some of the more elementary aspects of this theory will be developed for two-dimensional systems with occasional comments on higher-dimensional problems.

The linear two-dimensional system, with constant coefficients,

$$x' = ax + by \atop y' = cx + dy, \qquad ' = \frac{d}{dt} \qquad (1.1)$$

can be solved explicitly by the techniques developed in Chapter 1. This can be regarded as the first approximation of the nonlinear system

$$x' = f(x, y) \atop y' = g(x, y) \qquad (1.2)$$

97

where $f(x, y)$ and $g(x, y)$ satisfy $f(0, 0) = g(0, 0) = 0$ and have continuous derivatives, which at the origin are labeled as

$$\frac{\partial f(0, 0)}{\partial x} = a, \qquad \frac{\partial f(0, 0)}{\partial y} = b, \qquad \frac{\partial g(0, 0)}{\partial x} = c, \qquad \frac{\partial g(0, 0)}{\partial y} = d.$$

We will see that very exact knowledge of the behavior of solutions of (1.1) can often give qualitative knowledge of the behavior of solutions of (1.2) near the origin. To avoid some complicated situations, we will always assume that $ad - bc \neq 0$ (that the Jacobian of the right-hand side of (1.2) is not zero). The assumption that f and g have continuous derivatives implies that if a set of initial conditions, $x(t_0) = \alpha$, $y(t_0) = \beta$, is added to the system (1.2), the existence of a unique solution follows. These basic matters are summarized in the following formal statement, which is a special case of a theorem to be proved in Chapter 3.

THEOREM 1.1

Let $f(x, y)$, $g(x, y)$ be continuously differentiable. Then there is a unique solution of the initial value problem

$$x' = f(x, y), \qquad ' = \frac{d}{dt}$$

$$y' = g(x, y)$$

$$x(t_0) = \alpha, \qquad y(t_0) = \beta,$$

valid on an interval $I = (t_0 - \gamma, t_0 + \gamma)$. If this solution is denoted by $x(t, \alpha, \beta)$, $y(t, \alpha, \beta)$, then for a fixed value of $t \in I$, $x(t, \alpha, \beta)$, $y(t, \alpha, \beta)$ are continuous functions of α and β.

The solution above is defined for all $t \in R$ in the case of (1.1), but for (1.2) we must make additional assumptions on f and g to guarantee that a solution exists for all $t \in R$. Rather than impose technical conditions that will guarantee this, for this chapter we make the explicit assumption that all solutions of (1.2) exist for all $t \in R$—a continuability hypothesis.

In this chapter we will need to discuss certain sets of points in the plane and their properties. We introduce the terminology here for the reader who may not have encountered it before. Let R^2 denote the plane, and let points P in the plane be denoted by the ordered pair (x, y)—the usual x-y coordinates. We adopt the norm of Section 4 of Chapter 1. If $P = (x, y)$, $\|P\| = |x| + |y|$. Let $G \subset R^2$. A point P of G is said to be an *interior point* if there is a number r such that $\{|Q| \, \|P - Q\| < r\} \subset G$. In words rather than symbols, there is a number (perhaps small) such that all of the points closer (in the sense of norm) to P than this number are also in G. For example, if G is the "inside" of the unit circle $G = \{(x, y)|x^2 + y^2 < 1\}$, then every point of G is an interior point. More technically, if $(x, y) \in G$, then there is an $\varepsilon > 0$ such that all points $(\bar{x}, \bar{y})$ with $|x - \bar{x}| + |y - \bar{y}| < \varepsilon$ also satisfy $(\bar{x}, \bar{y}) \in G$. If $G' = \{(x, y)|(x, y) \notin G\}$, called the *complement* of G, then some points, those with $x^2 + y^2 > 1$, will be interior points of G' and those with $x^2 + y^2 = 1$ will not. A set that consists entirely of interior points is said to be an *open set*.

The open set $\{Q| \|P - Q\| < r\}$, for any $r > 0$, is said to be a *neighborhood* of P. If P is a point such that every neighborhood of P contains an element of the set G, then P is said to be a *limit point* of G. Note that $P \in G$ is *not* required. Thus in the circle example, points that satisfy $x^2 + y^2 = 1$ are limit points of G (and of G'). A set G whose limit points are elements of G is said to be a *closed set*. A set G, union the set of limit points of G, is called the *closure* of G. Obviously, the closure of a set is a closed set. Finally, by $A\backslash B$ we mean the points of a set A that are not points of B. We will use these concepts in the remainder of the chapter. The reader meeting them for the first time should work some of the exercises to be certain that he or she understands the definitions.

EXERCISES

Let

$$A = \{(x, y)|x^2 + y^2 \leq 1\}$$

$$B = \{(x, y)|x^2 + y^2 < 1\} \cup \{(x, y)|x = 1\}$$

$$C = \left\{(x, y)| \|(x, y)\| < \frac{1}{2}\right\}$$

$$D = \{(x, y)|(x - 1)^2 + y^2 < 1\}$$

1. Sketch the sets A, B, C, D in the plane.

2. Which of the sets A, B, C, D are open sets? Closed sets?

3. Which of the sets $A \cap B$, $A \cap C$, and $A \cap D$ are open? Closed?

4. Describe the limit points of the sets in Exercises 1 and 3.

5. Is the set $B\backslash A$ open? Closed? What is the set of limit points for this set?

6. Repeat Exercise 5 for $D\backslash B$.

7. Given an example of a set that is neither open nor closed.

8. Describe a set that is both open and closed.

2. The Phase Plane

Points along a solution of (1.2) can be viewed as a triple in R^3, $(t, x(t), y(t))$—a path traced out in three dimensions, a time coordinate t, and a two-dimensional space coordinate (x, y). The absence of the independent variable t in the right-hand side of (1.2) makes another interpretation useful. Solutions may be regarded in the plane as a parametric curve given by $(x(t), y(t))$, with t as the parameter. This curve is simply the projection of the triple $(t, x(t), y(t))$ in three-dimensional space onto the plane of the space variables. The curve $(x(t), y(t))$ is called a *trajectory* or an *orbit* and the plane is called the *phase plane*. This section explores the basis of this highly geometric approach with a view toward the applications that will come in later sections.

To see why the phase plane is a useful concept, we note first an elementary property of solutions of (1.2).

LEMMA 2.1

If $(\varphi_1(t), \varphi_2(t))$ is a solution of (1.2), so is $(\varphi_1(t - \tau), \varphi_2(t - \tau))$ for any real number τ.

Proof.   Define $\psi_1(t) = \varphi_1(t - \tau)$, $\psi_2(t) = \varphi_2(t - \tau)$. Then

$$\psi_1'(t) = \varphi_1'(t - \tau) = f(\varphi_1(t - \tau), \varphi_2(t - \tau))$$

$$= f(\psi_1(t), \psi_2(t)),$$

$$\psi_2'(t) = \varphi_2'(t - \tau) = g(\varphi_1(t - \tau), \varphi_2(t - \tau))$$

$$= g(\psi_1(t), \psi_2(t)),$$

and therefore $(\psi_1(t), \psi_2(t))$ solves (1.2).

Note that $(\varphi_1(t), \varphi_2(t))$, $t \in R$ and $(\varphi_1(t - \tau), \varphi_2(t - \tau))$, $t \in R$ describe the same set of points in the plane and hence the same trajectory.

We know that if solutions are viewed as points in R^3—as $(t, x(t), y(t))$, representing time and two space coordinates—there is a unique solution through each point. If we project these solutions onto the phase plane by using only $x(t)$, $y(t)$ as coordinates, might not a tangle of curves result? That this is not the case when t does not appear explicitly in f and g is the content of Theorem 2.2.

THEOREM 2.2

Let f and g be continuously differentiable. Through each point (x_0, y_0) of the plane there is a unique trajectory of

$$x' = f(x, y)$$
$$y' = g(x, y).$$

Proof. Suppose, to the contrary, there are two different trajectories $(\varphi_1(t),$ $\varphi_2(t))$, $(\psi_1(t), \psi_2(t))$ passing through (x_0, y_0), that is, $\varphi_1(t_0) = x_0 = \psi_1(t_1)$, $\varphi_2(t_0) = y_0 = \psi_2(t_1)$, where necessarily $t_0 \neq t_1$ (by the uniqueness of solutions of initial value problems). By Lemma 2.1, the functions

$$x_1(t) = \varphi_1(t - t_1 + t_0) \quad \text{and} \quad x_2(t) = \varphi_2(t - t_1 + t_0)$$

form a solution of (1.2). Yet

$$x_1(t_1) = \varphi_1(t_0) = x_0 = \psi_1(t_1)$$

and

$$x_2(t_1) = \varphi_2(t_0) = y_0 = \psi_2(t_1).$$

Then, by Theorem 1.1, the uniqueness theorem for initial value problems,

$$x_1(t) \equiv \psi_1(t) \quad \text{and} \quad x_2(t) \equiv \psi_2(t)$$

for all t, hence $(\varphi_1(t), \varphi_2(t))$ and $(\psi_1(t), \psi_2(t))$ are the same trajectories (with different parametrizations).

This same conclusion can be reached by considering that if $f(x_0, y_0) \neq 0$ in (1.2), then the initial value problem

$$\frac{dy}{dx} = \frac{g(x, y)}{f(x, y)}$$

$$y(x_0) = y_0$$

(2.1)

has a unique solution. Since

$$\frac{dy}{dx} = \frac{y'(t)}{x'(t)} = \frac{g(x(t), y(t))}{f(x(t), y(t))},$$

the solutions of (2.1) (locally) describe the trajectories of (1.2). This is a very useful way to describe, and sometimes to find, the curves that form the trajectories. We will use this method frequently in Section 3. Note that (2.1) is meaningful only if f and g are independent of the variable t. Differential equations for which the right-hand side does not contain t explicitly are said to be *autonomous*. If a point (x_0, y_0) is encountered on a trajectory where $f(x_0, y_0) = 0$, but $g(x_0, y_0) \neq 0$, we then consider

$$\frac{dx}{dy} = \frac{f(x, y)}{g(x, y)}$$

(2.2)

$$x(y_0) = x_0$$

instead. A point in the plane (x_0, y_0) such that

$$f(x_0, y_0) = 0$$
$$g(x_0, y_0) = 0$$

(2.3)

is said to be a *critical point* of (1.2). The preceding discussion does not apply in this case, but the constant functions $x(t) \equiv x_0$, $y(t) \equiv y_0$, form a solution of (1.2) and the entire trajectory is just the critical point. Again, using the uniqueness of solutions, it follows that *no trajectory may pass through a critical point*. (Other terms for critical point, in this context, are *equilibrium solution* or *stationary solution*.)

The function obtained by taking a point (x_0, y_0) into the value at time t of the (unique) solution of (1.2) with initial condition $x(0) = x_0$, $y(0) = y_0$ is called the *solution mapping*.

This idea is very important and we pause to explore it a little more. Recall that a mapping of the plane into itself is simply a function that takes a point in the plane into another point in the plane. Those functions met in calculus are usually of the form $z = f(x, y)$, $w = g(x, y)$, given explicitly by known functions f and g. For example, $z = -x$, $w = y$ takes each point (x, y) into its mirror image with respect to the y-axis. Consider now the system of differential equations

$$z' = Az$$

where

$$A = \begin{bmatrix} 0 & 1 \\ 1 & 0 \end{bmatrix}.$$

A fundamental matrix for this system is given by

$$e^{At} = \begin{bmatrix} \cosh(t) & \sinh(t) \\ \sinh(t) & \cosh(t) \end{bmatrix}.$$

We can describe the solution mapping for trajectories of this differential equation just as explicitly as we do the mirror-image mapping. The solution mapping takes the explicit form

$$T_t(x_0, y_0) = \begin{bmatrix} \cosh(t) & \sinh(t) \\ \sinh(t) & \cosh(t) \end{bmatrix} \begin{bmatrix} x_0 \\ y_0 \end{bmatrix}$$

$$= \begin{bmatrix} x_0 \cosh(t) + y_0 \sinh(t) \\ x_0 \sinh(t) + y_0 \cosh(t) \end{bmatrix}.$$

The subscript t on T_1 signifies that we have, in fact, described a whole family of mappings, indexed by the real number t. Thus

$$T_1(x_0, y_0) = \begin{bmatrix} x_0 \cosh(1) + y_0 \sinh(1) \\ x_0 \sinh(1) + y_0 \cosh(1) \end{bmatrix}$$

describes one mapping of the plane into itself, while

$$T_2(x_0, y_0) = \begin{bmatrix} x_0 \cosh(2) + y_0 \sinh(2) \\ x_0 \sinh(2) + y_0 \cosh(2) \end{bmatrix}$$

is a different mapping of the plane into itself. For the solution mapping, the important point is that we need not know the mapping explicitly—the existence theorem, Theorem 1.1, says that this mapping is defined at each point for small enough real numbers t, and our basic continuability hypothesis says that it is defined for all t. Thus, we can define the solution mapping for the nonlinear system

$$x' = y$$
$$y' = \sin(x)$$

without knowing the explicit formulation for it, provided we know that it can be continued. This is often a convenient way to visualize solutions of differential equations—to think of points in the plane as "flowing along" the trajectories in the plane propelled by the solution mapping.

The particular mapping just described has the property that

$$T_1(T_1(x_0, y_0)) = T_2(x_0, y_0).$$

To check this, we compute

$$\begin{bmatrix} \cosh(1) & \sinh(1) \\ \sinh(1) & \cosh(1) \end{bmatrix} \begin{bmatrix} \cosh(1) & \sinh(1) \\ \sinh(1) & \cosh(1) \end{bmatrix} \begin{bmatrix} x_0 \\ y_0 \end{bmatrix}$$

$$= \begin{bmatrix} \cosh^2(1) + \sinh^2(1) & 2\cosh(1)\sinh(1) \\ 2\sinh(1)\cosh(1) & \sinh^2(1) + \cosh^2(1) \end{bmatrix} \begin{bmatrix} x_0 \\ y_0 \end{bmatrix}.$$

A conversion of hyperbolic functions to exponential functions or a quick consultation with a book of tables with identities of hyperbolic functions shows that this is equal to

$$\begin{bmatrix} \cosh(2) & \sinh(2) \\ \sinh(2) & \cosh(2) \end{bmatrix} \begin{bmatrix} x_0 \\ y_0 \end{bmatrix} = T_2(x_0, y_0).$$

This point is intuitively clear, for it should make no difference whether we begin at (x_0, y_0) and flow along the (unique) trajectory through this point for one unit of time, then begin at the new point (new initial condition) and flow along the (unique) trajectory through it for one more unit of time, or simply go from the original point for two units of time. This property holds for all solution mappings for (1.2) under our hypotheses. We express the general case as

$$T_t(T_s(x_0, y_0)) = T_{t+s}(x_0, y_0).$$

The final sentence in Theorem 1.1 says that for a fixed t, $T_t(x, y)$ is continuous in the pair (x, y) in the usual sense of continuity of functions of two variables. T_t is also continuous in t.

In Section 9 of Chapter 1, we investigated the stability of the trivial solution. The following material duplicates the definitions and some of the material given there (since the reader may have omitted that chapter). The focus here, however, is on the two-dimensional case and allows for nonlinear differential equations.

Our interest centers on the behavior of solutions near—in a neighborhood of—a critical point. A critical point (x_0, y_0) of (1.2) is said to be *stable* if for every $\varepsilon > 0$ there is a $\delta > 0$ such that if any solution $(x(t), y(t))$ of (1.2) satisfies $|x(t_0) - x_0| + |y(t_0) - y_0| < \delta$, then it also satisfies $|x(t) - x_0| + |y(t) - y_0| < \varepsilon$ for all $t \geq t_0$. Roughly speaking, a critical point is stable if trajectories can be made to remain arbitrarily close to it for all future time by being sufficiently close to it at the present time. Critical points that are not stable are said to be *unstable*. If a critical point (x_0, y_0) is stable and if there is an $\eta > 0$ such that if any solution of (1.2) that satisfies $|x(t_0) - x_0| + |y(t_0) - y_0| < \eta$ also satisfies $\lim_{t \to \infty} |x(t) - x_0| + |y(t) - y_0| = 0$, then the critical point is said to be *asymptotically stable*. Again, roughly speaking, if a stable critical point has the additional property that solutions that come sufficiently close to it actually converge to it as $t \to \infty$, then it is asymptotically stable. For applications, it is asymptotic stability that is important, for if the initial conditions are slightly moved (perturbed) from the critical point, the trajectory through these initial conditions will have the critical point as its limiting value.

The ε's and δ's in the above definition make it clear that these concepts are local. System (1.2) may have several critical points, some stable, some unstable, and some asymptotically stable. The linear system (1.1) has only the origin as a critical point and it is the case that asymptotic stability, if it occurs, is a global property.

To understand the behavior of trajectories near a critical point, it is sometimes helpful to convert the system to a system of differential equations in variables $r(t), \theta(t)$, which yields the trajectory in polar coordinates. We do this for system (1.1). Let $(\varphi_1(t), \varphi_2(t))$ be a solution of (1.1). Define

$$r(t) = (\varphi_1^2(t) + \varphi_2^2(t))^{1/2}$$

$$\theta(t) = \tan^{-1}\left(\frac{\varphi_2(t)}{\varphi_1(t)}\right)$$

(2.4)

for $\varphi_1(t) \neq 0$, and by continuity elsewhere (see Exercise 11). Sometimes the transformation (2.4) will yield information directly, but sometimes it is helpful to have the set of differential equations for these quantities. Since

$$\varphi_1(t) = r(t)\cos\theta(t),$$

$$\varphi_2(t) = r(t)\sin\theta(t)$$

and they satisfy (1.1), it follows that

$$\varphi_1'(t) = r'(t)\cos\theta(t) - r(t)\sin\theta(t)\theta'(t)$$
$$= ar(t)\cos\theta(t) + br(t)\sin\theta(t),$$
$$\varphi_2'(t) = r'(t)\sin\theta(t) + r(t)\cos\theta(t)\theta'(t)$$
$$= cr(t)\cos\theta(t) + dr(t)\sin\theta(t).$$

Hence $r(t)$, $\theta(t)$ are solutions of the system

$$r' = r[a\cos^2\theta + d\sin^2\theta + (b+c)\sin\theta\cos\theta]$$

$$\theta' = c\cos^2\theta - b\sin^2\theta + (d-a)\sin\theta\cos\theta.$$

(2.5)

Note that $r(t) > 0$ if $r(0) > 0$, and that the right-hand side of the equation for θ' is independent of r. Conversely, given a solution $\rho(t)$, $\omega(t)$ of (2.5), $\varphi_1(t) = \rho(t)\cos\omega(t)$ and $\varphi_2(t) = \rho(t)\sin\omega(t)$ is a solution of (1.1). The solutions of (2.5) will sometimes directly yield information about the system (1.1). For example, if we can show that $\lim_{t\to\infty} r(t) = 0$, then that solution must tend to the origin. If $\lim_{t\to\infty} \theta(t) = \infty$, then $\varphi_2(t_n) = 0$ for a sequence $\{t_n\}$, $t_n \to \infty$. In this case, it is not difficult to argue that there is also a sequence $\{\tau_n\}$, $\tau_n \to \infty$ such that $\varphi_1(\tau_n) = 0$.

Finally, it should be remarked that the transformation (2.4) could be applied to nonlinear differential equations or to nonautonomous differential equations we well, and sometimes useful information can be obtained.

It will turn out (see Section 4) that the behavior of solutions of systems of the form (1.2) in a neighborhood of a critical point is related to the behavior of a system of the form (1.1). This system can be completely analyzed—in fact, explicitly solved, if necessary.

The definitions of critical point, stability, and so on carry over to n-dimensional nonlinear systems. Consider the system

$$y' = f(y)$$

where $y \in R^n$ and $f: R^n \to R^n$. A critical point is a vector y_0, $y_0 \in R^n$ such that $f(y_0) = 0$, where, of course, 0 represents the origin in R^n. More explicitly, if the above system is written componentwise,

$$y'_1 = f_1(y_1, \ldots, y_n)$$
$$y'_2 = f_2(y_1, \ldots, y_n)$$
$$\vdots$$
$$y'_n = f_n(y_1, \ldots, y_n),$$

then $y_0 = (y_1^0, y_2^0, \ldots, y_n^0)$ is a critical point if

$$f_i(y_1^0, y_2^0, \ldots, y_n^0) = 0, \qquad i = 1, \ldots, n.$$

Stability and asymptotic stability can be formulated using any convenient norm for R^n. Our choice is

$$\|y\| = \sum_{i=1}^{n} |y_i|$$

where $y = (y_1, \ldots, y_n)$.

EXERCISES

1. Find the differential equations for the polar functions r, θ of the following two-dimensional systems.

   (a) $x' = x + y$
   $y' = x - y$

   (b) $x' = y$
   $y' = -x$

   (c) $x' = x + y$
   $y' = -x + y$

   (d) $x' = x + 2y$
   $y' = x + 4y$

2. Convert the scalar equation $y'' + y = 0$ to a system and determine the polar functions. Is the origin stable? Asymptotically stable?

3. Convert the scalar system $y'' + \sin(y) = 0$ to a system and determine the differential equations for the corresponding r and θ.

4. Apply the transformation (2.4) to the general nonlinear system (1.2) and obtain differential equations for r and θ.

5. Define stability and asymptotic stability for critical points of an n-dimensional system of ordinary differential equations by analogy with the definitions for two-dimensional systems.

6. Locate the critical points of the following systems.

(a) $x' = y$
$\ y' = \sin(x)$

(b) $x' = x - y$
$\ y' = y^2 - x$

(c) $x' = x^2(y - 1)$
$\ y' = xy$

(d) $x' = x^2(y - 1)$
$\ y' = x^2 - 2xy - y^2$

(e) $x' = x - y^2$
$\ y' = x^2 - y^2$

(f) $x' = \sin(y)$
$\ y' = \cos(x)$

7. Find the polar coordinate equations for (a), (c), and (d) in Exercise 6.

8. Let Γ be the trajectory through (x_0, y_0) at $t = 0$. If (x_1, y_1) is a point of Γ, show that $T_t(x_0, y_0) = (x_1, y_1)$ for some value of t.

9. Show that $T_{t+s}(x_0, y_0) = T_t[T_s(x_0, y_0)]$. (*Hint:* Exercise 8.)

10. A fixed point of a mapping $T: R^2 \to R^2$ is a point (x, y) such that $T(x, y) = (x, y)$. Show that if the solution map T_1 has a fixed point, there is a periodic solution of (1.2) of period 1, that is, a solution that satisfies $x(t + 1) = x(t)$, $y(t + 1) = y(t)$.

In Exercises 11–13, let $(\phi_1(t), \phi_2(t))$ be a solution of (1.1) with initial conditions at $t = 0$, $\phi_2(t) > 0$, and $\phi_1(0) > 0$. Define $\theta_1(t)$ by (2.4), where the principal branch is chosen as $(-\pi/2 < \theta_1 < \pi/2)$, and suppose $\lim_{t \to t_0} \phi_1(t) = 0$ where t_0 is the first point to the right of zero where this occurs.

11. Compute $\lim_{t \to t_0^-} \theta_1'(t)$.

12. For $t_0 < t < t_0 + \varepsilon$, ε small, define $\theta_2(t)$ by (2.4) with $\pi/2 < \theta_2 < 3\pi/2$. Compute $\lim_{t \to t_0^+} \theta_2'(t)$.

13. Let $\theta(t) = \begin{cases} \theta_1(t), & 0 \leq t < t_0 \\ \theta_2(t), & t_0 < t < t_0 + \varepsilon \\ \dfrac{\pi}{2}, & t = t_0. \end{cases}$

Show that $\theta(t)$ is continuously differentiable.

3. Critical Points of Some Special Linear Systems

We now want to use the phase plane techniques to examine some special linear systems. These systems will be in a form that makes the computation of the eigenvalues and the eigenvectors and the conversion to polar coordinates easy. Analysis of these simple systems provides guidelines as to what sorts of behavior are possible. We shall see that these systems are not so special after all—that the general linear system (with nonzero eigenvalues) may be transformed into one of the six types to be discussed. The classification proceeds by the nature of the eigenvalue (and eigenvectors if a double eigenvalue occurs). One representative example is analyzed for each possible case.

Case 1 *The eigenvalues of A are real, distinct, and of the same sign.*

Take, as a representative, $A = \begin{pmatrix} \lambda & 0 \\ 0 & \mu \end{pmatrix}$. System (1.1) then is

$$x' = \lambda x$$
$$y' = \mu y,$$

which can be solved to obtain

$$x(t) = x_0 e^{\lambda t}$$
$$y(t) = y_0 e^{\mu t}.$$

If λ and μ are negative, $\lim_{t \to \infty} x(t) = 0$ and $\lim_{t \to \infty} y(t) = 0$. Since the convergence then is monotone, the origin is an asymptotically stable critical point. If λ, μ are positive, then $\lim_{t \to \infty} x(t) = \pm \infty$ and $\lim_{t \to \infty} y(t) = \pm \infty$. Since the limiting behavior is the same no matter how close the initial conditions are to the origin, this is sufficient to show that the origin is unstable. For nonlinear systems, we are interested only in behavior near the critical point and such detailed global behavior will not generally be known, so the following idea is useful. The instability of the origin follows from that fact that $\lim_{t \to -\infty} x(t) = 0$ and the fact that $\lim_{t \to -\infty} y(t) = 0$. The trajectory tends to the origin as time "runs backward." This is a convenient trick, so we pause to explain it a bit more. Consider

$$x' = f(x, y)$$
$$y' = g(x, y).$$

If we make a change of variable $\tau = -t$, the system becomes

$$-\frac{dx}{d\tau} = f(x, y)$$

$$-\frac{dy}{d\tau} = g(x, y).$$

The signs of all of the derivatives are reversed, but

$$\frac{dy}{dx} = \frac{g(x, y)}{f(x, y)},$$

so the differential equation of the trajectories is the same for both systems. (See equation (2.1) of the previous section.) The curves are the same but the parameterization is reversed. This is what we mean by "time running backward." Intuitively, if backward time "attracts," then forward time must "repel." Hence, the critical point is unstable. If either $x_0 = 0$ or $y_0 = 0$, the corresponding component remains zero for all time.

Since

$$\theta(t) = \tan^{-1}\left(\frac{y(t)}{x(t)}\right) = \tan^{-1}\left(\frac{y_0 e^{\mu t}}{x_0 e^{\lambda t}}\right)$$

$$= \tan^{-1}\left(\frac{y_0}{x_0}\right) e^{(\mu - \lambda)t},$$

then $\lim_{t \to \infty} \theta(t) = 0$ if $\lambda > \mu$, and $\lim_{t \to \infty} \theta(t) = \pm \pi/2$ if $\mu > \lambda$, except for solutions corresponding to $x_0 = 0$ or $y_0 = 0$, respectively. Limits as $t \to -\infty$ are reversed. The distinction between λ and μ is arbitrary; it merely determines how the axes are labeled. Behavior for $0 > \lambda > \mu$ is shown in Figure 3.1 and for $0 < \mu < \lambda$ in Figure 3.2. The arrows indicate the direction of increasing time. The

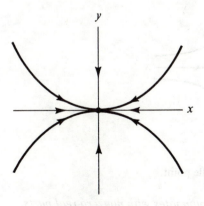

Figure 3.1 A stable node.

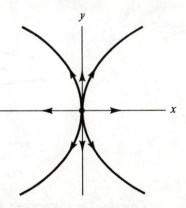

Figure 3.2 An unstable node.

solutions along the axes correspond to $x_0 = 0$ or $y_0 = 0$. In this case, the origin is said to be a *node*. It is important to note that, except for $x = 0$ or $y = 0$, the limiting angle, $\theta(t)$, is independent of the starting angle.

Case 2   *The eigenvalues are real and of opposite sign.*

The same representative matrix is selected as in the previous case, except that now it is assumed that $\lambda < 0 < \mu$. (The reverse case works similarly.) As before, solutions are $x(t) = x_0 e^{\lambda t}$, $y(t) = y_0 e^{\mu t}$ and $r(t) = (x_0^2 e^{2\lambda t} + y_0^2 e^{2\mu t})^{1/2}$. Since $\mu > 0$, if $y_0 \neq 0$, then $r(t)$ satisfies $\lim_{t \to \infty} r(t) = \infty$. Further, again if $y_0 \neq 0$,

$$\theta(t) = \tan^{-1}\left(\frac{y_0 e^{\mu t}}{x_0 e^{\lambda t}}\right)$$

satisfies $\lim_{t \to \infty} \theta(t) = \pm \pi/2$. If $y_0 = 0$, then $\lim_{t \to \infty} r(t) = 0$, and the trajectory approaches the origin with $\theta(t) = 0$ for all t. In this case, the origin is said to be a *saddle point*; Figure 3.3 illustrates the behavior of trajectories. The equation

$$\frac{dy}{dx} = \frac{y'}{x'} = \frac{\mu y}{\lambda x},$$

which is the equation of the trajectories (see (2.1)), can be solved to yield

$$yx^{-\mu/\lambda} = c.$$

This gives the equation of the "hyperbolic-looking" curves in Figure 3.3.

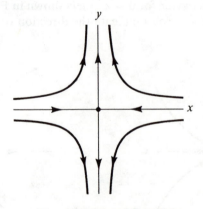

Figure 3.3   A saddle point.

Case 3   *The eigenvalues are complex conjugates with nonzero real parts.*

Take as a representative member of this class

$$A = \begin{bmatrix} \alpha & \beta \\ -\beta & \alpha \end{bmatrix}, \qquad \alpha\beta \neq 0,$$

so that the eigenvalues are $\lambda = \alpha \pm \beta i$ where, without loss of generality, we take $\beta > 0$. The system (1.1) is

$$x' = \alpha x + \beta y$$
$$y' = -\beta x + \alpha y.$$

Here polar coordinates are especially useful. The transformation (2.4) to polar functions yields

$$r' = \alpha r$$
$$\theta' = -\beta$$

where we have used (2.5) with $a = d = \alpha, b = -c = \beta$. This system may be solved easily, showing that all solutions are given by

$$r(t) = r_0 e^{\alpha t}$$
$$\theta(t) = \theta_0 - \beta t.$$

As $t \to \infty$, $\theta(t) \to -\infty$, so that solutions wind around the origin arbitrarily many times. The polar radius tends to zero as $t \to \infty$ if α is negative and, in this case, the critical point is asymptotically stable. The polar radius tends monotonically to $+\infty$ as $t \to \infty$ and to zero as $t \to -\infty$ if α is positive. Hence, in this case, the critical point is unstable. The shape of the curve can be easily obtained in polar coordinates. Since

$$\frac{dr}{d\theta} = \frac{r'}{\theta'} = \frac{\alpha r}{-\beta},$$

it follows that $\log r/r_0 = (\alpha/-\beta)(\theta - \theta_0)$, or that

$$r = r_0 e^{-(\alpha/-\beta)(\theta - \theta_0)}.$$

This curve is a logarithmic spiral, so the trajectory in the phase plane is a logarithmic spiral. This type of critical point is called a *spiral point*. (Other terms that are used are *focal point* or *vortex point*.) Figure 3.4 shows the case $\alpha < 0$.

Case 4   *The eigenvalues are purely imaginary.*

This is the same as the previous case but with $\alpha = 0$. The corresponding representative member of the class is

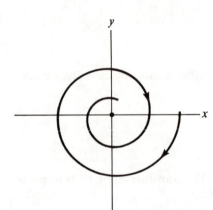

Figure 3.4 A stable spiral point.

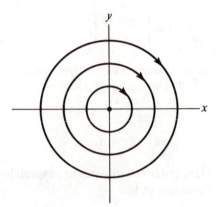

Figure 3.5 A center.

$$A = \begin{bmatrix} 0 & \beta \\ -\beta & 0 \end{bmatrix}.$$

The equations for the polar functions are very simple,

$$r' = 0$$
$$\theta' = -\beta,$$

and may be solved to obtain

$$r = r_0$$
$$\theta = -\beta t + \theta_0.$$

The trajectories are circles of radius r_0 about the critical point, as shown in Figure 3.5. This type of critical point is said to be a *center*. Since the trajectories in the phase plane are closed curves, the corresponding solutions are periodic. Since trajectories—circles—that begin near the origin remain there, the center is stable, but is not asymptotically stable.

Case 5 *The eigenvalues are coincident.*

Since the eigenvalues are equal, they are necessarily real. For this case, there are two possible representative elements (depending on whether there are one or two linearly independent eigenvectors corresponding to the repeated eigenvalue). First consider

$$A = \begin{pmatrix} \lambda & 0 \\ 0 & \lambda \end{pmatrix}.$$

System (1.1) then is

$$x' = \lambda x$$
$$y' = \lambda y,$$

and the equations are "uncoupled"—not related. A solution is

$$x(t) = x_0 e^{\lambda t}$$
$$y(t) = y_0 e^{\lambda t}.$$

Thus, $r(t) = (x^2(t) + y^2(t))^{1/2} = e^{\lambda t}(x_0^2 + y_0^2)^{1/2}$, and if $\lambda < 0$, $\lim_{t \to \infty} r(t) = 0$ (asymptotic stability) and if $\lambda > 0$, $\lim_{t \to \infty} r(t) = +\infty$. The polar angle is

$$\theta(t) = \tan^{-1}\left(\frac{y_0 e^{\lambda t}}{x_0 e^{\lambda t}}\right) = \tan^{-1}\left(\frac{y_0}{x_0}\right) = \theta_0$$

and the direction is constant, that is, trajectories are a half ray approaching or leaving the origin. Solutions are depicted in Figure 3.6 for $\lambda < 0$.

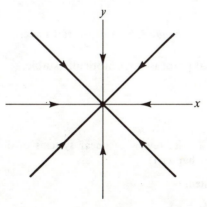

Figure 3.6   A degenerate node.

Such a critical point is called a *degenerate node*. In contrast to the behavior of Case 1 with negative real eigenvalues, there is no "preferred" direction of approach to the critical point.

Another possible representative for the case of coincident roots is

$$A = \begin{bmatrix} \lambda & 0 \\ 1 & \lambda \end{bmatrix}.$$

System (1.1) now becomes

$$x' = \lambda x$$

$$y' = x + \lambda y$$

with solution $x(t) = x_0 e^{\lambda t}, y(t) = y_0 e^{\lambda t} + x_0 t e^{\lambda t}$. If $\lambda < 0$, clearly $\lim_{t \to \infty} x(t) = 0$, $\lim_{t \to \infty} y(t) = 0$ and if $\lambda > 0$, $\lim_{t \to \infty} x(t) = +\infty$, $\lim_{t \to \infty} y(t) = +\infty$. If $x_0 = 0$, then the trajectory is half of the y-axis. If $x_0 \neq 0$, then $\theta(t) = \tan^{-1}((y_0 + x_0 t)/x_0)$ and $\lim_{t \to \infty} \theta(t) = \pi/2$ or $3\pi/2$. Figure 3.7 depicts the case $\lambda < 0$.

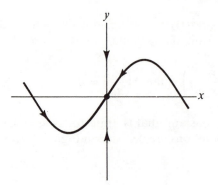

Figure 3.7   A degenerate node.

With $\lambda < 0$, the critical point is asymptotically stable.

EXERCISES

Exercises 1–7 provide a brief review of linear systems of differential equations (for the reader who skipped Chapter 1).

1. Show that the system

$$x' = ax + by$$

$$y' = cx + dy$$

has a solution of the form $e^{\lambda t} \begin{pmatrix} c_1 \\ c_2 \end{pmatrix}$ if and only if λ is an eigenvalue and $\begin{pmatrix} c_1 \\ c_2 \end{pmatrix}$ is an eigenvector of the matrix $A = \begin{pmatrix} a & b \\ c & d \end{pmatrix}$.

Two solutions (φ_1, φ_2) and (ψ_1, ψ_2) of (2.1) are said to be linearly dependent if there exist constants C_1, C_2, not both zero, such that $C_1 \begin{pmatrix} \varphi_1(t) \\ \varphi_2(t) \end{pmatrix} + C_2 \begin{pmatrix} \psi_1(t) \\ \psi_2(t) \end{pmatrix} = 0$

for $t \in R$. Solution pairs that are not linearly dependent are said to be linearly independent.

2. Show that if $\begin{pmatrix} \varphi_1 \\ \varphi_2 \end{pmatrix}$ and $\begin{pmatrix} \psi_1 \\ \psi_2 \end{pmatrix}$ are linearly independent solutions of (1.1), then the matrix

$$\Phi(t_0) = \begin{pmatrix} \varphi_1(t_0) & \psi_1(t_0) \\ \varphi_2(t_0) & \psi_2(t_0) \end{pmatrix}$$

is nonsingular.

3. Show that if $\begin{pmatrix} \varphi_1 \\ \varphi_2 \end{pmatrix}, \begin{pmatrix} \psi_1 \\ \psi_2 \end{pmatrix}$ are linearly independent solutions of (1.1), then every solution is a linear combination of these two solutions, that is, show that

$$\begin{pmatrix} \chi_1(t) \\ \chi_2(t) \end{pmatrix} = c_1 \begin{pmatrix} \varphi_1(t) \\ \varphi_2(t) \end{pmatrix} + c_2 \begin{pmatrix} \psi_1(t) \\ \psi_2(t) \end{pmatrix}$$

for every solution $\begin{pmatrix} \chi_1(t) \\ \chi_2(t) \end{pmatrix}$ of (1.1).

4. Use Exercises 1–3 to find all solutions of $\begin{pmatrix} x' \\ y' \end{pmatrix} = A \begin{pmatrix} x \\ y \end{pmatrix}$ where $A =$

(a) $\begin{pmatrix} 1 & 0 \\ 0 & 2 \end{pmatrix}$ (b) $\begin{pmatrix} 0 & 2 \\ 2 & 3 \end{pmatrix}$

(c) $\begin{pmatrix} 1 & -1 \\ -1 & 1 \end{pmatrix}$ (d) $\begin{pmatrix} 0 & 1 \\ 1 & -1 \end{pmatrix}$

5. Show that if the eigenvalues λ of A are complex conjugates, then $\mathrm{Re}(e^{\lambda t}c)$ and $\mathrm{Im}(e^{\lambda t}c)$ (real and imaginary parts of $e^{\lambda t}c$) are real solutions of (1.1). (*Hint:* Recall Euler's formula, $e^{i\theta} = \cos\theta + i\sin\theta$.)

6. Find solutions of $\begin{pmatrix} x' \\ y' \end{pmatrix} = A \begin{pmatrix} x \\ y \end{pmatrix}$ where $A =$

(a) $\begin{pmatrix} 0 & 1 \\ -1 & 0 \end{pmatrix}$ (b) $\begin{pmatrix} 1 & 1 \\ -1 & 1 \end{pmatrix}$

7. Find solutions of $\begin{pmatrix} x \\ y \end{pmatrix}' = A \begin{pmatrix} x \\ y \end{pmatrix}$ where $A =$

(a) $\begin{pmatrix} 1 & 0 \\ 0 & 1 \end{pmatrix}$ (b) $\begin{pmatrix} 2 & 1 \\ 1 & 0 \end{pmatrix}$ (c) $\begin{pmatrix} 0 & 2 \\ 2 & 4 \end{pmatrix}$

8. Determine the nature of the critical point at the origin of $\begin{pmatrix} x \\ y \end{pmatrix}' = A \begin{pmatrix} x \\ y \end{pmatrix}$ where $A =$

(a) $\begin{bmatrix} 2 & 0 \\ 0 & 3 \end{bmatrix}$ (b) $\begin{bmatrix} 2 & 3 \\ -3 & 2 \end{bmatrix}$

(c) $\begin{bmatrix} 4 & 1 \\ 0 & 4 \end{bmatrix}$ (d) $\begin{bmatrix} 1 & 0 \\ 0 & -4 \end{bmatrix}$

(e) $\begin{bmatrix} 0 & 7 \\ 7 & 0 \end{bmatrix}$ (f) $\begin{bmatrix} 2 & 0 \\ 0 & 2 \end{bmatrix}$

9. Find the polar equations for each system in Exercise 8.

10. Sketch the trajectories of the systems in Exercise 8.

11. Show formally that a center is stable but not asymptotically stable.

12. Show formally that if $\lim_{t \to \infty} r(t) = 0$ for (1.1), then the origin is asymptotically stable.

4. Critical Points of General Two-Dimensional Linear Systems

Computations with matrices are easier to follow if the matrices are simple. It turns out that the simple systems considered in the previous section do in fact comprise all possible behaviors for two-dimensional linear systems of differential equations (when the origin is an isolated critical point). The general system can be transformed into one of the simple systems of Section 3, making computations easier. To do this, we change the coordinate system where the equation is defined, or, equivalently, introduce a new dependent variable that transforms system (1.1) into a more convenient form. However, the reader may wish to ignore this bit of theory, treat the cases described in the preceding section as examples rather than a complete characterization, and move on to the Section 5.

Let B be a nonsingular 2×2 matrix and define a vector (z, w) by

$$\begin{pmatrix} z \\ w \end{pmatrix}(t) = B \begin{pmatrix} x \\ y \end{pmatrix}(t). \tag{4.1}$$

Then, if $(x(t), y(t))$ is a solution of (1.1),

$$\begin{pmatrix} z'(t) \\ w'(t) \end{pmatrix} = B \begin{pmatrix} x'(t) \\ y'(t) \end{pmatrix} = BA \begin{pmatrix} x(t) \\ y(t) \end{pmatrix}$$

$$= BAB^{-1} \begin{pmatrix} z(t) \\ w(t) \end{pmatrix},$$

or, defining $T = BAB^{-1}$, the vector (z, w) solves the differential equation

$$\begin{pmatrix} z' \\ w' \end{pmatrix} = T \begin{pmatrix} z \\ w \end{pmatrix}. \tag{4.2}$$

Since $T = BAB^{-1}$, T is *similar* to A. Similar matrices have the same eigenvalues, and since our classification will depend on the eigenvalues of A, we may work with a simple element of the set of matrices that are similar to A. In Section 3, we enumerated six cases that exhaust the possibilities for the eigenvalues of the matrix $A = \begin{pmatrix} a & b \\ c & d \end{pmatrix}$. It is possible to show that the matrix T in (4.2) is one of these six types. (Note that the assumption $ad - bc \neq 0$ precludes a zero eigenvalue and guarantees that $(0, 0)$ is the only critical point of system (1.1).)

The importance of this observation is that all linear systems will behave in the same way as the representative discussed in Section 3, where "essentially" means up to whatever distortion the change of variables (4.1) allows. The distortion is simply multiplication by a constant matrix, and it is possible to prove linear algebra theorems that describe this change in detail. Instead of developing this theory, we shall content ourselves with looking at a typical example.

Before introducing the example, however, we note first what is not changed by (4.1). If trajectories of the system (4.2) (analyzed in Section 3) tend to the origin as $\lim_{t \to \infty}$ or as $\lim_{t \to -\infty}$, then so will those of (4.1). This can be made clear by the use of norms, discussed in Section 4 of Chapter 1. Suppose we analyze (4.2) and find that

$$\lim_{t \to \infty} [z(t), w(t)] = 0.$$

Then,

$$\| (x(t), y(t)) \| = \| B^{-1} (z(t), w(t))^T \|$$
$$\leq \| B^{-1} \| \, \| (z(t), w(t)) \|.$$

Thus, if $\| (z(t), w(t)) \|$ tends to zero as t tends to infinity, so does $\| (x(t), y(t)) \|$, since $\| B^{-1} \|$ is just a fixed constant. Hence, $\lim_{t \to \infty} (x(t), y(t)) = 0$. (Of course, the roles of (z, w) and (x, y) can be interchanged, using B instead of B^{-1}.)

What can change are qualitative properties such as the direction of approach to the origin, the nice circles in the case of a center, or very hyperbolic-looking saddles. The following example treats one case of a saddle point and is representative of what can happen. The notion of "qualitative property" will be made more precise in the next section.

Consider the system

$$x' = ax + by$$
$$y' = cx + dy$$

where $\lambda < 0 < \mu$ are the eigenvalues of

$$A = \begin{bmatrix} a & b \\ c & d \end{bmatrix}.$$

Let $\begin{pmatrix} c_1 \\ c_2 \end{pmatrix}$ be an eigenvector corresponding to λ, let $\begin{pmatrix} d_1 \\ d_2 \end{pmatrix}$ be an eigenvector corresponding to μ, and suppose, for convenience, that c_1 and d_1 are not zero. Then a fundamental matrix is

$$\Phi(t) = \begin{pmatrix} c_1 e^{\lambda t} & d_1 e^{\mu t} \\ c_2 e^{\lambda t} & d_2 e^{\mu t} \end{pmatrix},$$

and since every solution can be represented as

$$\Phi(t) \begin{pmatrix} K_1 \\ K_2 \end{pmatrix},$$

an arbitrary solution takes the form

$$x(t) = \begin{pmatrix} K_1 c_1 e^{\lambda t} + K_2 d_1 e^{\mu t} \\ K_1 c_2 e^{\lambda t} + K_2 d_2 e^{\mu t} \end{pmatrix}.$$

The polar angle $\theta(t)$ corresponding to $x(t)$ satisfies

$$\tan \theta(t) = \frac{K_1 c_2 e^{\lambda t} + K_2 d_2 e^{\mu t}}{K_1 c_1 e^{\lambda t} + K_2 d_1 e^{\mu t}}$$

$$= \frac{K_1 c_2 e^{(\lambda - \mu)t} + K_2 d_2}{K_1 c_1 e^{(\lambda - \mu)t} + K_2 d_1}.$$

Hence

$$\lim_{t \to \infty} \tan \theta(t) = \frac{d_2}{d_1}.$$

This is the "slope" of the eigenvector $\begin{pmatrix} d_1 \\ d_2 \end{pmatrix}$; or, *solutions have the limiting direction as the direction of an eigenvector.* Similarly,

$$\tan \theta(t) = \frac{K_1 c_2 + K_2 d_2 e^{(\mu - \lambda)t}}{K_1 c_1 + K_2 d_1 e^{(\mu - \lambda)t}}$$

and

$$\lim_{t \to -\infty} \tan \theta(t) = \frac{c_2}{c_1},$$

the direction of the other eigenvector. (In the representative form, the eigenvectors were $\binom{0}{1}$ and $\binom{1}{0}$.) For example, the system

$$x' = \begin{pmatrix} 1 & 1 \\ 4 & 1 \end{pmatrix} x$$

has eigenvalues $\lambda = -1$, $\mu = 3$, and corresponding eigenvectors

$$\begin{pmatrix} c_1 \\ c_2 \end{pmatrix} = \begin{pmatrix} -1 \\ 2 \end{pmatrix}, \qquad \begin{pmatrix} d_1 \\ d_2 \end{pmatrix} = \begin{pmatrix} 1 \\ 2 \end{pmatrix}.$$

A phase plane plot is shown in Figure 4.1.

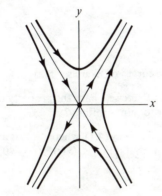

Figure 4.1   A saddle point before transformation.

A similar analysis can be made for the various nodes with fixed directions of approach.

EXERCISES

1. Let $A = [a_{ij}]$ and $T = [t_{ij}]$ for $i, j = 1, 2$, and denote the columns of T by t^1 and t^2, that is,

$$t^i = \begin{pmatrix} t_{1i} \\ t_{2i} \end{pmatrix}.$$

Show by direct computation that $AT = [At^1, At^2]$ where At^i is intended to be a column vector within the matrix.

2. Rework Exercise 1, using the summation notation for matrix multiplication (i.e., $AT = [\sum_k a_{ik} t_{kj}]$) and matrix-vector multiplication.

3. Use the formalism of Exercise 2 to extend the result to the n-dimensional case.

4. Let A be similar to a diagonal matrix. Show that if $D = T^{-1}AT$, D diagonal, then the columns of T are eigenvectors of A. (*Hint:* $AT = TD$; try to visualize multiplication on the right by a diagonal matrix.)

5. Change the system $x' = Ax$ to diagonal form by a suitable transformation, where A is

(a) $\begin{bmatrix} 1 & 3 \\ 3 & 2 \end{bmatrix}$ (b) $\begin{bmatrix} -1 & 1 \\ 1 & 2 \end{bmatrix}$

(c) $\begin{bmatrix} -1 & -3 \\ -3 & -1 \end{bmatrix}$ (d) $\begin{bmatrix} 0 & -1 \\ -1 & 0 \end{bmatrix}$

(e) $\begin{bmatrix} 0 & -1 & 3 \\ -1 & 1 & 0 \\ 2 & 0 & 1 \end{bmatrix}$

(*Hint:* Use Exercise 4.)

6. Sketch the phase plane curves of Exercises 5(a)–(d).

7. Which of the systems in Exercise 5 have the property that all solutions tend to zero as t tends to infinity or to minus infinity?

8. Let (z, w) and (x, y) be related by (4.1), where (x, y) is a solution of (1.1) and (z, w) is a solution of (4.2). Show that if $\lim_{t\to\infty} (x(t), y(t)) = 0$, then $\lim_{t\to\infty} (z(t), w(t)) = 0$.

5. Behavior of Nonlinear Two-Dimensional Systems Near a Critical Point

We turn now to the main problem of interest, the system

$$x' = f(x, y)$$
$$y' = g(x, y), \tag{5.1}$$

where it is assumed that f and g have continuous partial derivatives—an assumption continued for the remainder of the chapter. A consequence of this assumption is that solutions of the initial value problem (5.1) with $x(t_0) = x_0$, $y(t_0) = y_0$ are unique. It is also assumed that solutions of (5.1) exist for $-\infty < t < \infty$, although this is seldom critical in the analysis.

We first put our equation into a more convenient form by the use of Taylor's theorem. Let (x_0, y_0) be an (isolated) critical point of (5.1); that is, assume $f(x_0, y_0) = g(x_0, y_0) = 0$ and $f_x(x_0, y_0)g_y(x_0, y_0) - f_y(x_0, y_0)g_x(x_0, y_0) \neq 0$, where h_z denotes $\partial h/\partial z$. Then, by Taylor's theorem, (5.1) may be written as

$$x' = f_x(x_0, y_0)(x - x_0) + f_y(x_0, y_0)(y - y_0) + \varepsilon_1(x - x_0, y - y_0)$$
$$y' = g_x(x_0, y_0)(x - x_0) + g_y(x_0, y_0)(y - y_0) + \varepsilon_2(x - x_0, y - y_0). \tag{5.2}$$

If (x_0, y_0) is translated to the origin by $z_1 = x - x_0$, $z_2 = y - y_0$, and if we write $a = f_x(x_0, y_0)$, $b = f_y(x_0, y_0)$, $c = g_x(x_0, y_0)$, $d = g_y(x_0, y_0)$, then (5.2) may be written as

$$z_1' = az_1 + bz_2 + \varepsilon_1(z_1, z_2)$$
$$z_2' = cz_1 + dz_2 + \varepsilon_2(z_1, z_2). \tag{5.3}$$

The linear part of (5.3) is the same as the system (1.1) studied in Section 4, $\varepsilon_i(z_1, z_2)$ is continuously differentiable, $\varepsilon_i(0, 0) = 0$, and $\partial\varepsilon_i/\partial z_j(0, 0) = 0$. This says that $\varepsilon_i(z_1, z_2)$, $i = 1, 2$ contains no constant or linear terms.

Before making the relationship between (5.3) and (1.1) clear, we pause to note some of their differences. System (5.1) may have many critical points; one has been selected and moved to the origin. If a different critical point is chosen, the constants a, b, c, d in the transformed equation (5.3) may be different, and, of course, the resulting behavior may be different. The point is that any statements that relate (1.1) to (5.3) are local in nature in that they apply near—in a neighborhood of—the critical point. Totally different kinds of behavior may occur in a neighborhood of other critical points. The transformation above must be made, and the linear part analyzed, for each isolated critical point of the nonlinear system.

Stability and asymptotic stability were defined as local properties, something that occurred in a neighborhood of a critical point, although in all of the *linear* examples it was, in fact, a global property. We now come to one of the reasons why it was defined this way. The asymptotic stability of (5.3) in the neighborhood of the origin is determined by that of (1.1). This is progress, because we now have a necessary and sufficient condition for the asymptotic stability of the origin for the system (1.1)—that all of the eigenvalues have negative real parts. The following theorem makes the relationship between the stabilities of (1.1) and (5.3) precise.

THEOREM 5.1

If the origin is an asymptotically stable critical point for (1.1) then it is an asymptotically stable critical point for (5.3). If the origin is unstable for (1.1), then it is unstable for (5.3).

We illustrate the theorem with an example. Take, for (5.3), the system

$$x' = -x + y - x(y - x)$$
$$y' = -x - y + 2x^2 y \tag{5.4}$$

and take for (1.1) the linear system

$$x' = -x + y$$
$$y' = -x - y. \tag{5.5}$$

The coefficient matrix for the linear equation (5.5) is

$$\begin{bmatrix} -1 & 1 \\ -1 & -1 \end{bmatrix},$$

which has eigenvalues

$$\lambda = -1 \pm i.$$

Since (5.5) has the origin as an asymptotically stable critical point, the theorem states that the nonlinear system (5.4) does also. All solutions of (5.5) tend to the origin but not all solutions of (5.4) do. For example, $(1, 1)$ is another critical point for (5.4). However, by Theorem 5.1, all solutions of (5.4) that start sufficiently close to the origin tend to the origin. How close is "sufficiently close" is generally a difficult question to answer. Behavior near the critical point $(1, 1)$ must be determined separately.

More than this can be said, but to do so requires a new definition. Before introducing the new concept, we note, for use in future examples, that the polar equations for (5.3) take the form

$$r' = r[a\cos^2(\theta) + d\sin^2(\theta) + (b + c)\sin(\theta)\cos(\theta)]$$
$$+ \varepsilon_1 \cos(\theta) + \varepsilon_2 \sin(\theta),$$
$$\theta' = c\cos^2(\theta) - b\sin^2(\theta) + (d - a)\sin(\theta)\cos(\theta) \tag{5.6}$$
$$+ \frac{\varepsilon_2 \cos(\theta) - \varepsilon_1 \sin(\theta)}{r}.$$

Two autonomous systems

$$x' = f(x, y)$$
$$y' = g(x, y)$$

(5.7)

and

$$u' = \alpha(u, v)$$
$$v' = \beta(u, v)$$

(5.8)

are said to have the same *qualitative structure* in regions G_1 and G_2 of the appropriate planes if there is a one-to-one mapping T that takes G_1 onto G_2 such that

1. T and T^{-1} are continuous;

2. If two points of G_1 lie on the same trajectory of (5.7), their images under T lie on the same trajectory of (5.8);

3. If two points of G_2 lie on the same trajectory of (5.8), their images under T^{-1} lie on the same trajectory of (5.7).

The mapping T in the definition may, at first, appear quite mysterious. Condition (1) makes this mapping "nice" in many ways that we will not be able to explore here—so nice, and so important, that topologists give this kind of mapping a name, *homeomorphism*. The mapping is shown schematically in Figure 5.1. Conditions (2) and (3) require the mapping and its inverse to preserve trajectories. So far, no hint has been given as to when there might be such a mapping.

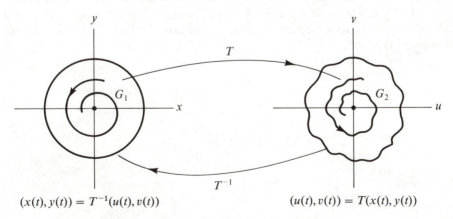

$$(x(t), y(t)) = T^{-1}(u(t), v(t)) \qquad\qquad (u(t), v(t)) = T(x(t), y(t))$$

Figure 5.1 A schematic of the mapping T.

The use of this concept is to make precise the "sameness" of the two different systems of differential equations. T can be viewed as a change of coordinates under which trajectories of one of the systems become trajectories of the other. However, having the same qualitative structure does not imply having the same asymptotic stability properties, as may be seen by considering the following example. For the two systems, take

$$x' = x$$
$$y' = y$$

and

$$x' = -x$$
$$y' = -y.$$

For the mapping T, take the identity mapping—the function that sends each point into itself. The trajectories in the phase plane of these two systems are exactly the same, so the mapping T fulfills all of the conditions necessary for the two systems to have the same qualitative structure. However, one set of trajectories tends to the origin as t tends to infinity, while the other set tends to the origin as t tends to minus infinity. The curves are the same in the phase plane, solutions just run along them in opposite directions. This will not be a problem for us, however, since it turns out that one *sufficient* condition for two systems to have the same qualitative structure in a neighborhood of a critical point is also a sufficient condition for the two critical points to have the same asymptotic stability properties. Before stating this theorem, however, we look at an example. Although the concept is valuable for relating solutions of nonlinear differential equations, the construction of such a mapping and the identification of proper domains G_1 and G_2 is difficult. For linear systems, we have effectively done that already in the previous section in selecting the canonical forms, so the example presented is a linear one.

Consider the two systems

$$y' = \begin{bmatrix} -1 & -3 \\ -3 & -1 \end{bmatrix} y \tag{5.9}$$

and

$$x' = \begin{bmatrix} 1 & 0 \\ 0 & -2 \end{bmatrix} x, \tag{5.10}$$

where x and y are column vectors with entries $x_i, y_i, i = 1, 2$. The trajectories are

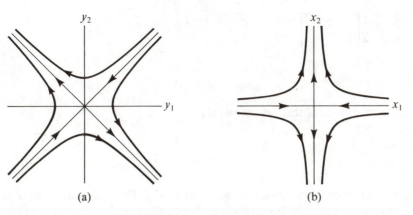

Figure 5.2 Trajectories of systems (5.9) and (5.10).

sketched in Figure 5.2. The equations for the trajectories of (5.10) are given by the solutions of

$$\frac{dx_2}{dx_1} = -\frac{2x_2}{x_1}.$$ (5.11)

Take G_1 and G_2 to be the entire (respective) planes and let the mapping T be defined by

$$T\begin{pmatrix} y_1 \\ y_2 \end{pmatrix} = \begin{pmatrix} y_1 - y_2 \\ y_1 + y_2 \end{pmatrix} = \begin{bmatrix} 1 & -1 \\ 1 & 1 \end{bmatrix}\begin{bmatrix} y_1 \\ y_2 \end{bmatrix}.$$ (5.12)

Clearly, T is continuous and its inverse, given by

$$T^{-1}\begin{pmatrix} z_1 \\ z_2 \end{pmatrix} = \frac{1}{2}\begin{pmatrix} z_1 + z_2 \\ -z_1 + z_2 \end{pmatrix},$$ (5.13)

is continuous. To see that trajectories of (5.9) are mapped onto trajectories of (5.10), we find the differential equation satisfied by Ty when $y(t)$ is a solution of (5.9). Let $z(t)$ be a function defined by $z = Ty$. Then

$$z' = (Ty)' = Ty'$$

$$= \begin{bmatrix} 1 & -1 \\ 1 & 1 \end{bmatrix} y'$$

$$= \begin{bmatrix} 1 & -1 \\ 1 & 1 \end{bmatrix}\begin{bmatrix} -1 & -3 \\ -3 & -1 \end{bmatrix} y$$

$$= \frac{1}{2} \begin{bmatrix} 1 & -1 \\ 1 & 1 \end{bmatrix} \begin{bmatrix} -1 & -3 \\ -3 & -1 \end{bmatrix} \begin{bmatrix} 1 & 1 \\ -1 & 1 \end{bmatrix} z$$

$$= \begin{bmatrix} 2 & 0 \\ 0 & -4 \end{bmatrix} z.$$

Hence, trajectories of this system satisfy

$$\frac{dz_2}{dz_1} = -\frac{2z_2}{z_1},$$

which is the same differential equation as (5.11). Thus, the mapping T maps trajectories of (5.9) onto trajectories of (5.10). We leave as an exercise that T^{-1} maps trajectories of (5.10) onto those of (5.9).

For nonlinear equations, the construction would not be so easy. The following theorem can be viewed as an existence theorem for the mapping T and for the regions G_1 and G_2 in a neighborhood of a critical point for systems (1.1) and (5.3).

THEOREM 5.2

If the real parts of the eigenvalues of $\begin{bmatrix} a & b \\ c & d \end{bmatrix}$ are not zero, (5.3) and (1.1) have the same qualitative structure in a neighborhood of the origin.

Thus if two systems have the same linear part, and the real parts of the eigenvalues are not zero, they have the same stability *and* the same qualitative structure. In particular, we can take (5.8) to be a nonlinear equation, while (5.7) is a linear equation, and since the behavior of the linear system is known, we have a "qualitative clue" as to the behavior of the nonlinear one. For example, the linear part of (5.3),

$$\xi_1' = a\xi_1 + b\xi_2$$
$$\xi_2' = c\xi_1 + d\xi_2$$

(which is the same as (1.1)), is called the *linearization* of (5.1) about the critical point (x_0, y_0). The content of Theorem 5.1 is that the stability of the system (5.2) near—and "how near" is not resolved—the critical point is determined by the stability of its linearization. Theorem 5.2 goes further in relating the "sameness" of the two equations. It tells us that in a small region around the critical point, solutions of (1.1) and its linearization may be mapped onto each other by a "nice" mapping—one that is continuous and has a continuous inverse. Thus, geometric properties of the systems that are preserved by such mappings are the same. Theorem 5.2 is a first step in applying topological methods to ordinary differential equations.

There is a particular curve, which will be of interest in Section 7, that we wish to identify. If the real parts of both eigenvalues of a linear system are positive, then we can find a closed curve about the origin that every trajectory of the canonical form system crosses "outward," that is, that every trajectory crosses from inside to outside the circle as time increases. If the real parts of both eigenvalues are negative, then a similar statement holds with the trajectories crossing inward. This is summarized in the following theorem (where we are being "fuzzy" about inside and outside—see Theorem 7.2 on page 144.

THEOREM 5.3

If the real parts of the eigenvalues of $\begin{bmatrix} a & b \\ c & d \end{bmatrix}$ are positive (negative), then there is a closed curve about the origin that trajectories of (5.3) cross from inside to outside (outside to inside) as time increases.

Two cautions are in order when we use the theorems presented here. First, the conclusions are all local—they say something about the behavior of the nonlinear equation near a critical point. It has been noted that a nonlinear differential equation may have several critical points and the theorems may be applied near each one. Nothing whatever is said about behavior away from the critical points. This is, in general, a difficult problem and solutions of nonlinear equations can behave differently from solutions of linear ones—for example, limit cycles, to be found in Section 7. Sometimes it may be possible, after analyzing the local behavior, to see a way to fill in the space away from the critical points. Such a conjecture motivates further analysis, either analytical or computational. The second caution is that the "nice" pictures of Section 2, "square" saddle points, logarithmic spirals, and so on, may be distorted by the transformation T of Theorem 5.2. The essential features remain, but the figures may be stretched, irregularly rotated, or bent. Under the assumptions we are making about f and g, we can define "spiral point" for (5.1) so as to make spiral points of the linear system correspond to spiral points of the nonlinear system. The critical point (x_0, y_0) of (5.1) is said to be a *spiral point* if, after translating (x_0, y_0) to $(0, 0)$ and changing to polar coordinates (as in (5.6)), it follows that $\lim_{t \to \infty} r(t) = 0$ and $\lim_{t \to \infty} \theta(t) = \pm \infty$ (or $\lim_{t \to -\infty}$ of the same quantities).

THEOREM 5.4

If the linearization of (5.1) about (x_0, y_0) has the origin as a spiral point, then (x_0, y_0) is a spiral point for (5.3).

The spirals for (5.3), of course, may be distorted and not the nice logarithmic spirals of (1.1) Similar definitions of nodes and saddles could be made for nonlinear two-dimensional systems so that counterparts of Theorem 5.4 hold.

To see how this approach works, consider

$$x' = y$$
$$y' = x - y + x(x - 2y). \tag{5.14}$$

The critical points, which are solutions of

$$y = 0$$
$$x - y + x^2 - 2xy = 0,$$

are $(0, 0)$ and $(-1, 0)$. Furthermore, $f_x = 0, f_y = 1, g_x = 1 + 2x - 2y$, and $g_y = -1 - 2x$. At $(0, 0)$ the matrix of the linear part is

$$\begin{pmatrix} 0 & 1 \\ 1 & -1 \end{pmatrix},$$

whose eigenvalues satisfy $\lambda^2 + \lambda - 1 = 0$. The roots are real and of opposite sign. The origin then is a saddle point for (5.14). At $(-1, 0)$ the matrix corresponding to the linear part is

$$\begin{pmatrix} 0 & 1 \\ -1 & 1 \end{pmatrix},$$

whose eigenvalues satisfy

$$\lambda^2 - \lambda + 1 = 0 \quad \text{or} \quad \lambda = \frac{1 \pm \sqrt{-3}}{2} = \frac{1}{2} \pm \frac{\sqrt{3}}{2}i.$$

This critical point is an unstable spiral for the system (5.14).

Figure 5.3 shows the trajectories of this system. It is easy to see how to fill in the phase plane away from the two critical points. Further analysis is needed, however, to justify this picture.

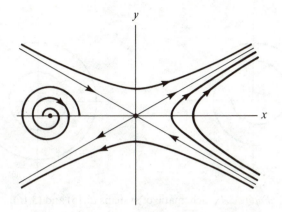

Figure 5.3   Trajectories of system (5.14).

The case where the linear system is a center—where the eigenvalues are purely imaginary—remains. In this case, the behavior of the linearization gives no information as to the behavior of the nonlinear system (which is why it has been excluded). The first example shows that a center for the linear system may be a spiral for the nonlinear system.

Consider

$$x_1' = -x_2 - x_1(x_1^2 + x_2^2)$$
$$x_2' = x_1 - x_2(x_1^2 + x_2^2). \tag{5.15}$$

The origin is a critical point and the linearization about it is given by

$$\begin{pmatrix} x_1 \\ x_2 \end{pmatrix}' = \begin{pmatrix} 0 & -1 \\ 1 & 0 \end{pmatrix} \begin{pmatrix} x_1 \\ x_2 \end{pmatrix}, \tag{5.16}$$

so that it is a center. The polar equations (5.6) are (after simplification)

$$r' = -r^3$$
$$\theta' = 1.$$

Clearly, $\theta(t) = \theta_0 + t$ and $\lim_{t\to\infty} \theta(t) = \infty$. The equation for r may be solved to obtain $r(t) = (2t + (1/r_0^2))^{-1/2}$, and hence $\lim_{t\to\infty} r(t) = 0$. Clearly, trajectories approach the origin in a spiral fashion. See Figure 5.4. (Note that it is not a logarithmic spiral.) Changing signs of the nonlinear part would produce an unstable spiral. (Note that solutions of (5.15) are defined only for $t > -1/2r_0^2$. As $t \to -1/2r_0^2$, $r(t) \to \infty$. This is a phenomenon of nonlinear differential equations for which we must always be alert.)

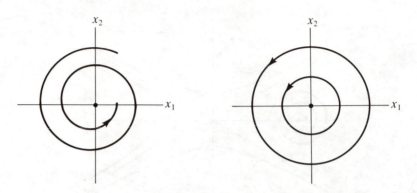

Figure 5.4   Schematic of systems (5.15) and (5.16).

To see how complicated a center may be, consider the following system,

$$x' = -y + x(x^2 + y^2)\sin\left(\frac{\pi}{\sqrt{x^2 + y^2}}\right)$$
$$\quad , \quad ' = \frac{d}{dt} \quad\quad (5.17)$$
$$y' = x + y(x^2 + y^2)\sin\left(\frac{\pi}{\sqrt{x^2 + y^2}}\right)$$

which has polar equations (see (5.6))

$$r' = r^3 \sin\left(\frac{\pi}{r}\right)$$
$$\theta' = 1.$$

The quantity $r = 1/n$, $\theta = t$ is a solution of the polar equations (since the derivative of a constant is zero) yielding, as a trajectory, a circle of radius $1/n$. Note that if $r > 1$, then $r' > 0$. Furthermore, if $1/(2n + 1) < r < 1/2n$, then $r' > 0$, but if $1/2n < r < 1/(2n - 1)$, $r' < 0$, $n = 1, 2, \ldots$. If $r(0) \neq 1/n$, then r is monotonic, so as $t \to \infty$ or as $t \to -\infty$ the trajectory spirals onto one of the circles (see Figure 5.5). The closed curves correspond to "isolated" periodic solutions and there is a "countably infinite" number of them. Thus, the structure of the nonlinear system is very complicated.

EXERCISES

1. Locate the critical points for the following nonlinear systems.

    (a) $x' = y$
           $y' = -\sin(x)$

    (b) $x' = x + x^2$
           $y' = x + y^2$

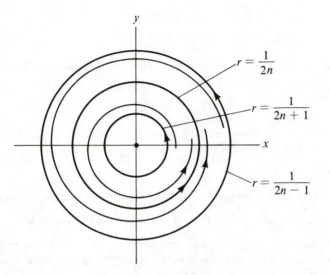

Figure 5.5 A portion of a complicated center (system 5.17).

(c) $x' = x + y - 2y^2$
$y' = -x + y - 2y^2$

(d) $x' = x - x^2 - \lambda_1 xy$
$y' = y - y^2 - \lambda_2 xy$

(e) $x' = x(3 - x - y)$
$y' = y(x - y - 1)$

(f) $x' = 1 - x$
$y' = xy - x$

(g) $x' = y$
$y' = -x + x^3$

(h) $x' = x(1 - y)$
$y' = y(1 - 2x)$

2. Determine the stability of the systems in Exercise 1(d) as a function of λ_1 and λ_2.

3. Derive equation (5.6).

4. Show that the functions ε_i in (5.3) are differentiable and that the first partial derivatives are zero at the origin. (*Hint:* ε_i is the difference of two differentiable functions.)

5. Suppose that two nonlinear systems

$$x' = f(x, y)$$
$$y' = g(x, y)$$

$$(*)$$

and

$$u' = h(u, v)$$
$$v' = k(u, v)$$

$$(**)$$

each have the same qualitative structure as the linear system

$$r' = ar + bs$$
$$s' = cr + ds \tag{***}$$

in regions G_1 (for (*) and (**)) and G_2, respectively. Show that (*) and (**) have the same qualitative structure in G_1. (*Hint:* Construct the required mapping as a composite.)

6. Show that if f and g have continuous partial derivatives, then x'' and y'' exist where $x(t)$ and $y(t)$ are solutions of (5.1). (*Hint:* Apply the definition of derivative to x' or, for a shortcut, use the chain rule.)

7. Verify Figure 5.3.

8. Verify that the mappings defined by (5.12) and (5.13) are inverses of each other.

9. Show that T^{-1}, given in (5.13), maps trajectories of (5.10) onto trajectories of (5.9).

10. Show that the property of being a closed trajectory is invariant under the identifying mapping of two systems of differential equations that have the same qualitative structure.

11. Show that a trajectory corresponds to a periodic solution if and only if it is a closed path parameterized by time.

6. Elementary Liapunov Stability Theory

The preceding sections analyzed the behavior of solutions of autonomous second-order differential equations in the neighborhood of a critical point. This analysis provided a local theory—it described solutions that become "close" to constant solutions, but did not give a method for determining which solutions become "close" to these constant solutions. Moreover, it is not only critical points that exhibit stability properties. The Liapunov theory, which is sketched in this section, provides a way to determine the global stability of certain sets of points. Before considering the theory, let us illustrate the basic idea with a simple example.

Consider a one-parameter family of circles

$$x^2 + y^2 = C$$

and a two-dimensional system of differential equations

$$x' = -x + y^2$$
$$y' = -y - xy. \tag{6.1}$$

Every choice of the constant $C \geq 0$ selects one of the circles ($C = 0$ selects the degenerate circle, the origin) and each point in the plane lies on exactly one of the circles. Instead of asking for the coordinates of a given trajectory as time increases, we ask only "on which circle does the trajectory lie at time t?" Let $\Gamma(t) = (x(t), y(t))$ be a given solution of (6.1) with $\Gamma(t_0) = \gamma_0$. Then at each time t, $\Gamma(t)$ lies on the circle determined by the number

$$x^2(t) + y^2(t).$$

Thus we can define a function $C(t)$ to be this quantity, the square of the radius of the circle on which $\Gamma(t)$ lies at any time t. We can then ask how $C(t)$ changes with t. If this is a less difficult problem, we are making progress. It turns out to be an easy problem, since

$$\frac{dC}{dt} = \frac{d}{dt}(x^2(t) + y^2(t))$$

$$= 2x(t)x'(t) + 2y(t)y'(t)$$

$$= 2x(t)(-x(t) + y^2(t)) + 2y(t)(-y(t) - x(t)y(t))$$

$$= -2x^2(t) - 2y^2(t),$$

where the fact that $(x(t), y(t))$ satisfies (6.1) has been used to replace terms $x'(t)$ and $y'(t)$. Hence

$$C'(t) = -2C(t)$$

or

$$C(t) = C_0 e^{-2(t-t_0)}, \quad C_0 = x^2(t_0) + y^2(t_0)$$

so that

$$\lim_{t \to \infty} C(t) = 0.$$

Therefore,

$$\lim_{t \to \infty} [x^2(t) + y^2(t)] = 0.$$

The trajectory $\Gamma(t)$ must approach the origin as $t \to \infty$. There is nothing in the limiting argument that depends on the choice of the particular trajectory $\Gamma(t)$, so the argument can be repeated for each orbit (only the constant C_0 might change).

Thus, every trajectory of (6.1) tends to the origin as $t \to \infty$. The origin is a *globally* asymptotically stable critical point.

If the system (6.1) were

$$x' = -\tfrac{1}{2}x + y^2$$
$$y' = -y - xy,$$

(6.2)

then for the same function C, we would have

$$C' = -x^2(t) - 2y^2(t).$$

The above argument would not continue, but we do have

$$C'(t) \leq -x^2(t) - y^2(t) = -C(t).$$

This inequality may be multiplied by e^t, rewritten as

$$(C(t)e^t)' \leq 0,$$

and integrated to obtain

$$C(t) \leq C_0 e^{-(t-t_0)}.$$

(We have "solved" the inequality.) The same conclusion follows—the origin is globally, asymptotically stable.

Geometrically, the idea is very simple. View the equation $C = x^2 + y^2$ as a surface in E^3, as shown in Figure 6.1. Level surfaces, $C = $ constant, are the circles

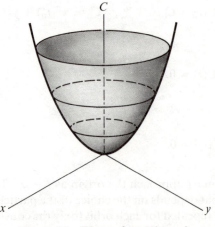

Figure 6.1   The surface $V(x, y, c)$.

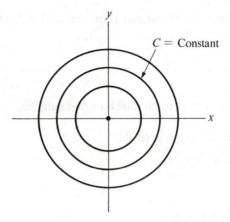

Figure 6.2   Level curves of the surface in Figure 6.1.

in Figure 6.2. Fix a solution $(x(t), y(t))$. Then, at time t, there is a corresponding point on the surface given by $(x(t), y(t), C(t))$. As time increases, the point moves "down the surface" in the sense that it moves to a lower "level." There is no "stopping point" except the point $(0, 0, 0)$, which is approached in the limit. In terms of the level curves, this means that the trajectory crosses circles with ever smaller radii as time increases and cannot stop on a circle with a positive radius. (It "slows down," of course, as the trajectory approaches the origin.)

Before developing this idea, we pause to note that the derivative used in the argument above can be expressed in a simple way. Let $V(x, y)$ be a function that maps R^2 into R and let ∇V be the vector $(V_x(x, y), V_y(x, y))^T$—the gradient. If x and y are functions of a parameter t, then the chain rule takes the form

$$\frac{d}{dt} V(x(t), y(t)) = \nabla V \cdot (x'(t), y'(t))^T$$

$$= V_x x' + V_y y'.$$

Further, if $x(t)$ and $y(t)$ are solutions of the differential equation

$$x' = f(x, y)$$
$$y' = g(x, y),$$

then

$$\frac{dV}{dt} = V_x f(x, y) + V_y g(x, y).$$

This quantity is abbreviated $\dot{V}$. It is instructive to keep the example above with the circle in mind when interpreting the following results.

We can now formalize the above argument and make it independent of the particular function $x^2 + y^2$.

THEOREM 6.1

Let $V(x, y)$ be a differentiable function that satisfies

    i. $V(x, y) > 0, \quad (x, y) \neq (0, 0),$

   ii. $V(0, 0) = 0$

  iii. $\dot{V} \leq kV$

for some constant $k < 0$.
Then every solution of

$$x' = f(x, y)$$
$$y' = g(x, y)$$

satisfies

$$\lim_{t \to \infty} x(t) = 0$$

$$\lim_{t \to \infty} y(t) = 0.$$

This is a weak theorem, since the hypotheses on $V(x, y)$ are very strong, particularly (iii). In spite of this, the theorem does find application in cases where the prospect of finding an explicit solution are slight. Consider the system

$$x' = y - x$$
$$y' = -y - x^3$$

and the function $V(x, y) = (1/2)y^2 + (1/4)x^4$. Then $\nabla V = (x^3, y)^T$. Hence

$$\dot{V} = x^3 y - x^4 - x^3 y - y^2$$
$$= -x^4 - y^2$$
$$\leq -\frac{x^4}{4} - \frac{y^2}{2} = -V.$$

Theorem 6.1 applies, and all solutions tend to zero as t tends to infinity.

The more general Liapunov theorem presented next, provides a way to make a portion of the above procedure work under far less restrictive circumstances. The interested reader should consult the References (particularly LaSalle) for a more complete understanding of the wide scope of the method.

Let

$$x' = f(x), \qquad x \in R^n \tag{6.3}$$

be an n-dimensional system of differential equations. Let $f(x)$ be defined on G^*, an open set in R^n, and let G be a subset of G^*. A function $V(x) : G \subset R^n \to R$ is said to be a *Liapunov function* for (6.3) on G if

1. V is continuously differentiable at each point $x \in G$, and

2. $\dot{V} = \nabla V \cdot f \leq 0$ on G.

The power of the Liapunov method is that $\dot{V}$ can be computed without a knowledge of solutions of the differential equation. In the first example, $n = 2$, $G^* = G = R^2$, and $V = x^2 + y^2$. For a general Liapunov function the set $\{x \in R^n \mid V(x) = C\}$ is a level surface, and these surfaces replace the circles $x^2 + y^2 = C$ used in the first example. They may be quite arbitrary (see Figure 6.3). The condition that V be continuously differentiable is restrictive and can be easily removed in the theory, but the resulting computations are correspondingly more difficult to carry out in specific problems.

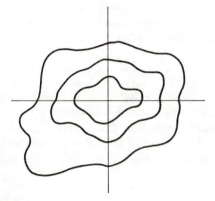

Figure 6.3   Schematic of level curves of an arbitrary Liapunov function.

In order to describe the sets of points that can be "stable," it is necessary to introduce several definitions. First, a set Ω is said to be *positively invariant* with respect to (6.3) if the trajectory through $x_0 \in \Omega$ at $t = 0$ remains in Ω for all $t \geq 0$.

Stated another way, Ω is positively invariant if $T_t(\Omega) \subset \Omega$ for all $t \geq 0$, where T_t is the solution map discussed in Section 2. *Negatively invariant* is defined by replacing $t \geq 0$ by $t \leq 0$. A set is *invariant* if it is positively and negatively invariant. (Thus $T_t(\Omega) = \Omega$ for $t \in R$.) Critical points, for example, are positively and negatively invariant; the set of points on a periodic orbit is also invariant. However, invariant sets can be quite complicated. The *distance* between a point p and a closed set Δ, both in R^n, is defined by

$$d(p, \Delta) = \min_x [\|(x - p)\| \,|\, x \in \Delta]$$

where $\|\ \ \|$ is the usual norm (see Section 4, Chapter 1). Recall that $\Gamma(t) = (x_1(t), \ldots, x_n(t))$ is said to approach a set Δ if $\lim_{t \to \infty} d(\Gamma(t), \Delta) = 0$. That a trajectory may approach a set of points without approaching any specific point of the set is illustrated by the final example in Section 5 by the trajectories that spiral onto a periodic trajectory.

The principal result is contained in the following theorem, which is a special case of what is commonly called the LaSalle corollary to the Liapunov stability theorem. (Note that $\overline{G}$ means the closure of G.)

THEOREM 6.2 *(Liapunov-LaSalle)*
Let V be a Liapunov function for

$$x' = f(x) \tag{6.3}$$

on a region G. Let $E = \{x \,|\, \dot{V}(x) = 0, x \in \overline{G} \cap G^*\}$ and let M be the largest invariant set in E. Then every bounded (for $t \geq 0$) trajectory of (6.3) that remains in G tends to the set M as $t \to \infty$.

The proof of this theorem, which is well beyond the scope of this book, can be found in LaSalle in the references. We illustrate it with several examples.

In the first example, equation (6.1), take $V(x, y) = x^2 + y^2$ and $G = G^* = R^2$. Then, since $\dot{V} = -2x^2 - 2y^2$, $E = \{(0,0)\}$ and $M = E$. Hence all (bounded) solutions tend to the origin as $t \to \infty$ and the origin is asymptotically stable, by Theorem 6.2.

Consider

$$x' = -xy^2$$
$$y' = -x^4 y \tag{6.4}$$

and again let $V = x^2 + y^2$ and $G = G^* = R^2$. Then,

$$\dot{V} = 2xx' + 2yy'$$
$$= -2x^2y^2 - 2x^4y^2$$
$$= -2x^2y^2(1 + x^2) \le 0,$$

$E = \{(x, y)|x = 0 \text{ or } y = 0\}$, and all of E is invariant. Therefore, every bounded trajectory approaches a subset of the x- or the y-axis (or possibly both). Each axis consists entirely of critical points, and it is not difficult to show that a trajectory must approach a single critical point.

Consider the system

$$x' = -y$$
$$y' = -y + g(x)$$

where $xg(x) > 0$, $x \neq 0$. Let

$$V(x, y) = \frac{y^2}{2} + \int_0^x g(\zeta)\, d\zeta$$

and let $G = G^* = R^2$. Then

$$\dot{V} = yy' + g(x)x'$$
$$= -y^2 + yg(x) + g(x)(-y) = -y^2 \le 0.$$

Therefore, $E = \{(x, y)|y = 0\}$. Let $(\xi, 0) \in M$, the largest invariant set in E. Then, for a trajectory through $(\xi, 0)$ to remain in M, it must satisfy the system

$$x' = 0$$
$$y' = g(x) = 0,$$

since to remain in E it must be the case that $y(t) = y'(t) = 0$. By the first equation, $x'(t) = 0$ or $x(t) \equiv C$, a constant. Therefore, $g(C) = 0$, but the only value of x where $g(x) = 0$ is $x = 0$. Thus, the only invariant set in E is the point $(0, 0)$ and by the Liapunov theorem, all bounded solutions tend to the origin.

Sometimes the Liapunov theory can be used to determine the region of attraction for a critical point. Consider the system

$$x' = -y$$
$$y' = x + y^3 - y \tag{6.5}$$

The only critical point is the origin. Let

$$V(x, y) = x^2 + y^2$$

and compute

$$\dot{V} = -2xy + 2xy + 2y^4 - 2y^2$$
$$= 2y^2(y^2 - 1)$$

Therefore, $V(x, y)$ will be a Liapunov function for (6.5) if $|y| < 1$. Consider the interior of the level curve $x^2 + y^2 = 1$ (the largest level curve in the region $|y| \leq 1$) as the region G; that is, $G = \{(x, y)|x^2 + y^2 < 1\}$. Along any solution that starts in this region,

$$\dot{V}(x, y) < 0 \qquad \text{if } y \neq 0.$$

The set $E = \{(x, y)|y = 0 \text{ or } y = \pm 1\}$ and $M = \{(0, 0)\}$; so all solutions that start in G tend to the origin. Thus, G represents a set of points "attracted" to E under the solution mapping.

It is not always the case that the limiting set is a critical point, as we shall see in the next section. All of the examples above yield a conclusion about bounded solutions. It is possible, by strengthening the conditions in the definition of Liapunov function, to conclude boundedness, but we leave this topic for the reader to pursue.

EXERCISES

1. Use $V(x, y) = x^2 + y^2$ to analyze

   (a) $\quad x' = -x^3 + 2xy^2$
   $\quad\quad y' = -2x^2y - y^3$

   (b) $\quad x' = y - x^3$
   $\quad\quad y' = -x$

   (c) $\quad x' = -y$
   $\quad\quad y' = x + y^5 - 2y$

   in an appropriate region. What stability conclusions can be drawn?

2. Show that $V(x, y) = y^2/2 - \cos(x)$ is a Liapunov function for

   $$x' = y$$
   $$y' = -\sin(x).$$

   Explain why no asymptotic stability conclusion can be drawn.

3. (a) Analyze the scalar equation

   $$y'' + f(y)y' + h(y) = 0, \qquad\qquad (*)$$

where $f(y) > 0$ and $yh(y) > 0$, $y \neq 0$, and f and h are continuous, by converting to a system of differential equations and finding a suitable Liapunov function.

(b)   Show that if, in addition, $\lim_{|y| \to \infty} \int_0^y h(s) \, ds = +\infty$, then all solutions of $(*)$ are bounded.

4. Modify the results of Exercise 3 to apply to the damped pendulum equation

$$y'' + y' + \sin(y) = 0.$$

5. In addition to being a Liapunov function for

$$x' = f(x, y)$$

$$y' = g(x, y)$$

on $G = R^2$, suppose that (a) $V(x, y) \geq 0$ and (b) $\lim_{x^2 + y^2 \to \infty} V(x, y) = +\infty$. Show that all solutions of the system of differential equations are bounded for $t \geq 0$.

6. Show that every trajectory of (6.4) that is bounded (in the future) tends to a critical point as $t \to \infty$.

7.   Limit Cycles and the Poincaré-Bendixson Theorem

An exclusive phenomenon of nonlinear differential equations is that of a "limit cycle"—the trajectory of a periodic solution that is the "limit" (in a sense that we will make precise) of nearby nonperiodic trajectories. We begin with an example. Consider the two-dimensional system

$$x' = x + y - x(x^2 + y^2)$$
$$y' = -x + y - y(x^2 + y^2). \tag{7.1}$$

The linearization near the origin is given by

$$\begin{pmatrix} x \\ y \end{pmatrix}' = \begin{pmatrix} 1 & 1 \\ -1 & 1 \end{pmatrix} \begin{pmatrix} x \\ y \end{pmatrix},$$

and it follows that the origin is an unstable spiral point, since the eigenvalues of the coefficient matrix are complex with positive real parts. Thus, all trajectories near the origin spiral away from it. To better understand the behavior of trajectories of the full nonlinear system (7.1), convert that system to polar coordinates by letting $x = r \cos \theta$, $y = r \sin \theta$. System (7.1) then becomes

$$r' = r(1 - r^2)$$
$$\theta' = -1. \tag{7.2}$$

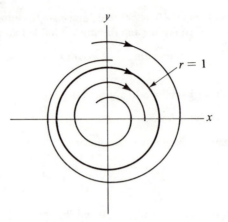

Figure 7.1 Phase plane plot for system (5.1).

The second equation can be solved to give $\theta(t) = \theta_0 - t$. If $r > 1$, then $r' < 0$, so trajectories spiral "in," while if $r < 1$, then $r' > 0$, and trajectories spiral "out." See Figure 7.1. The closed curve, $r = 1$, $\theta = -t$, corresponds to the trajectory of the periodic solution $x = \cos(t)$, $y = -\sin(t)$. (Other solutions that have the same orbit are $x = \cos(\theta_0 - t)$, $y = \sin(\theta_0 - t)$.) All orbits with initial condition $0 < r(0) \neq 1$ spiral to this closed orbit. If $\Gamma(t)$ is such a trajectory and C is the circle $r = 1$, then $\lim_{t\to\infty} d(\Gamma(t), C) = 0$, where d is the distance function defined in Section 6 (the distance between a point $\Gamma(t)$, for some t, and the closed set C). This behavior is a consequence of a famous mathematical theorem, the Poincaré-Bendixson theorem, one of the few general theorems for nonlinear differential equations. Before the theorem can be properly stated, some definitions and some notation must be introduced.

Let t_n be a sequence of real numbers such that $t_n \to \infty$ as $n \to \infty$. (Such a sequence is sometimes called an *extensive sequence*.) If $(x(t), y(t))$ is a solution of

$$\begin{aligned} x' &= f(x, y) \\ y' &= g(x, y) \end{aligned}, \qquad ' = \frac{d}{dt}, \qquad (7.3)$$

and if $P_n = (x(t_n), y(t_n))$ converges to a point P in the plane, then P is said to be an *omega limit point* of the trajectory $(x(t), y(t))$. The set of all such points is said to be the *omega limit set* of the trajectory $(x(t), y(t))$. The omega limit set may be a single point, as in the case of a globally, asymptotically stable critical point, or, as in the example above, it may be a closed curve. There are also more complicated possibilities, such as saddle points with connecting orbits. See Figure 7.2. However, if P is an omega limit point of a trajectory, it can be shown that *the entire trajectory through P consists of omega limit points*. In the language of the preceding section, the omega limit set is an invariant set.

One can examine the structure of omega limit sets in great detail using the

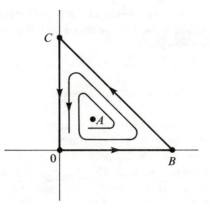

Figure 7.2   An example of a complex omega limit set. *A* is an unstable spiral point and *B*, *C*, and *O* are saddle points. A trajectory starting inside the triangle will have an omega limit set consisting of the critical points *O*, *B*, *C* and the orbits joining them, *OB*, *BC*, *CO*.

tools of topology. We give one example. Two nonempty sets of points, *A* and *B*, in the plane are said to be separated if there is no sequence of points p_n, $p_n \in A$ such that $\lim_{n \to \infty} p_n = q \in B$, and no sequence $q_n \in B$ such that $\lim_{n \to \infty} q_n = p \in A$. Two disjoint circles in the plane are separated; the interior of the positive octant and the *x*-axis are not. A set that is not the union of two separated sets is said to be *connected*. A *region* in the plane is an open, connected set. Although the definition is technical, sets that we would intuitively think of as connected in a nontechnical sense are connected. It can be shown that *the omega limit set of a bounded trajectory is a nonempty, closed, bounded, connected set.*

If $t_n \to -\infty$, we can similarly define an *alpha limit set* with the same properties. We now give the formal definition of limit cycle. Periodic orbits that are the omega or alpha limit sets of other orbits are called *limit cycles*. The nested circles in the example at the end of Section 5, corresponding to a complicated center, are alternately alpha and omega limit sets of the trajectories spiraling in between. Closed orbits correspond to periodic solutions, and in this example, each of the periodic solutions is a limit cycle. For each periodic orbit *C* there is a nonperiodic orbit, $\Gamma(t) = (x(t), y(t))$, such that $d(\Gamma(t), C) \to 0$ as $t \to \infty$ or $t \to -\infty$ (or both).

The principal result is the following.

THEOREM 7.1   (*Poincaré-Bendixson*)

   Let $\Gamma(t) = (x(t), y(t))$ be a trajectory of

$$x' = f(x, y)$$

$$y' = g(x, y), \qquad f, g \in C',$$

such that for $t \geq t_0$, $\Gamma(t)$ remains in a closed and bounded region of the plane, containing no critical points. Then, either $\Gamma(t)$ is a periodic orbit or its omega limit set is a periodic orbit.

This theorem requires advanced mathematical techniques to prove, and the proof will be omitted.

The Poincaré-Bendixson theorem is false for systems of dimension 3 or more. The proof depends heavily on a deep mathematical result, known as the Jordan curve theorem, which is valid only in the plane. Although the theorem sounds obvious, its proof is very difficult. We state the theorem for the convenience (and enlightenment) of the reader. We need first to be careful of the definition of a curve. Let $x = \varphi(t)$, $y = \psi(t)$, where $\varphi(t)$ and $\psi(t)$ are defined and continuous on an interval $t_0 \leq t \leq T$. If for any pair of points t_1, t_2, with $t_0 \leq t_1 < t_2 < T$, either $\varphi(t_1) \neq \varphi(t_2)$ or $\psi(t_1) \neq \psi(t_2)$, then the set of points defined by $(\varphi(t), \psi(t))$, $t_0 \leq t < T$ is said to be a *simple curve*. Further, if $\varphi(t_0) = \varphi(T)$ and $\psi(t_0) = \psi(T)$, the curve is *closed*.

THEOREM 7.2   *(Jordan curve theorem)*

A simple closed curve C divides the plane into two regions and is the boundary of these regions. One region (called the inside) is bounded and the other (called the outside) is unbounded.

Now that the inside of a curve is defined, the following result helps to clarify the type of region to seek in order to apply the Poincaré-Bendixson theorem.

THEOREM 7.3

Every periodic trajectory of

$$x' = f(x, y)$$

$$y' = g(x, y)$$

contains a critical point in its interior.

This theorem tells us that to apply the Poincaré-Bendixson theorem, the critical point-free region where the trajectory lies must have at least one hole in it—for the critical point. The example of system (7.1) provides a typical case for application. Consider a circle of radius $r_0 < 1$ about the origin. Since $r'(t) > 0$, all trajectories cross this circle "outward." Consider a circle of radius $r_1 > 1$. All trajectories cross this curve "inward." There are no critical points in the annulus $r_0 \le r \le r_1$ and every trajectory in this region at $t = t_0$ remains there for $t > t_0$, so this region must contain a periodic orbit. Note that nothing is said about uniqueness. The theorem can be applied to a critical point-free region, $r_0 \le r \le r_1$, of the system (5.9) in Section 5 with, say, $r_0 = 7/24$, $r_1 = 3/4$, to conclude the existence of a periodic orbit. In fact, there are two isolated periodic orbits (at $r = 1/2$, $r = 1/3$) in this annulus.

When the origin is unstable, Theorem 5.3 can be used to construct the curve around the critical point that all trajectories cross outward. The Liapunov theorem frequently can be used to obtain a bound on solutions. Together these results make an application of Theorem 7.1 possible. An example is given in the next section.

A set is said to be an *attractor* if it contains the omega limit set of the orbit through every point of a neighborhood of itself. A periodic trajectory can be an attractor, as shown in the example at the beginning of this section. The set of initial conditions where trajectories have the given set as an omega limit set is called the *basin of attraction*. If the basin of attraction is the entire set where the equations are being studied, the attractor is said to be a *global attractor*.

It is also important to know when there are *no* periodic orbits. The stability analysis of critical points is entirely local, but if it is known, in addition, that there are no periodic solutions, the omega limit set of a bounded trajectory must lie among the critical points (and, possibly, orbits connecting critical points). The absence of limit cycles can frequently be used to turn a local argument into a global one when we have an asymptotically stable critical point.

Unfortunately, general theorems that exclude periodic orbits are rare. We state and illustrate one such result that is frequently useful. First, a region G is said to be *simply connected* if for any simple closed curve C lying entirely within G, all points inside C are points of G.

THEOREM 7.4   (*Bendixson-Dulac criterion*)

Suppose there exists a continuously differentiable function $\beta(x, y)$ defined on a simply connected domain G. Suppose that the function

$$\frac{\partial}{\partial x}(\beta f) + \frac{\partial}{\partial y}(\beta g)$$

does not change sign in G. Then there are no periodic solutions of

$$x' = f(x, y)$$

$$y' = g(x, y)$$

in the region G.

A proof can be made using Green's theorem. While the proof is not difficult, presenting it would require additional material. The interested reader can find a proof in the references. We illustrate the theorem here; a more significant example will be given in the next section.

Consider the system

$$x' = y$$
$$y' = -x - (1 + x^2)y.$$

(The origin is a stable spiral.) Using $\beta \equiv 1$, $\partial f/\partial x = 0$, $\partial g/\partial y = -(1 + x^2)$, it follows that $\partial f/\partial x + \partial g/\partial y < 0$. Theorem 7.4 allows us to conclude that there are no periodic orbits in the plane.

Consider

$$x' = y$$
$$y' = -x - y + x^2 + y^2.$$

Choose $\beta(x, y) = e^{-2x}$, and note that

$$\frac{\partial}{\partial x}(ye^{-2x}) = -2ye^{-2x}$$

and

$$\frac{\partial}{\partial y}[(-x - y + x^2 + y^2)e^{-2x}] = -e^{-2x} + 2ye^{-2x}.$$

Thus,

$$\frac{\partial(\beta f)}{\partial x} + \frac{\partial(\beta g)}{\partial y} = -e^{-2x} < 0,$$

so by Theorem 7.4 there are no periodic solutions in the plane.

The Bendixson-Dulac criterion has much in common with the Liapunov theory in that "finding" the function $\beta(x, y)$ requires the same kind of ingenuity.

To see how limit cycles arise in two-dimensional systems, we modify the example of (7.1) to

$$x' = \alpha x + y - x(x^2 + y^2)$$
$$y' = -x + \alpha y - y(x^2 + y^2) \tag{7.4}$$

where α is a parameter. This means that we wish to consider an entire family of systems of differential equations, one system for each choice of α. Fixing α chooses a system of differential equations to which the previous theory may be applied. If $\alpha = 1$, we know from the analysis of (7.1) that there is a limit cycle. What happens when α changes? Does the new system of differential equations also have a limit cycle? In physical problems, the differential equations are derived from scientific principles and the parameter usually represents something that is fixed in each application but that varies between applications—for example, the length of a pendulum or the reproductive rate of a population.

As with (7.1), we apply the polar coordinate technique to change to r and θ variables. Thus (7.4) may be studied by analyzing

$$r' = r(\alpha - r^2)$$
$$\theta' = -1. \tag{7.5}$$

The behavior of trajectories of the system, as a function of α, is now transparent. If $\alpha < 0$, then $r' < \alpha r$ for any positive value of r and any value of θ, so $\lim_{t \to \infty} r(t) = 0$. The origin is (globally) asymptotically stable. If $\alpha > 0$, there is always a trajectory with $r = \sqrt{\alpha}$, $\theta = -t$ that corresponds to a periodic solution. In this case, if the initial conditions satisfy $r(0) < \sqrt{\alpha}$, then $r'(t) > 0$, and for an initial condition with $r(0) > \sqrt{\alpha}$, $r'(t) < 0$. Thus all nontrivial solutions tend to the limit cycle and the origin is an unstable critical point (a *repeller*). The change in behavior occurs at $\alpha = 0$. All of the systems with $\alpha < 0$ have the same qualitative structure as do all of those with $\alpha > 0$, but the two structures are different. At $\alpha = 0$, the system undergoes a dramatic change.

Let us explore the change a bit further. View α as decreasing. The radius of the circle representing the trajectory of the limit cycle decreases to zero coinciding with the origin at $\alpha = 0$. As α continues to decrease, there is no limit cycle and the origin is asymptotically stable. If we view α as increasing through zero from negative to positive, the critical point splits into a critical point and a limit cycle. Stability is "transferred" from the critical point to the limit cycle, since for $\alpha > 0$ the origin is a repeller. (We are being deliberately vague about the meaning of stability as applied to a limit cycle, but at least it is an attractor while the critical point has become a repeller.) The point $\alpha = 0$ is called a *bifurcation point*, and this particular phenomenon is called *Hopf bifurcation*. Its rigorous study requires considerable advanced mathematics.

EXERCISES

Exercises 1–3 are set-theoretic, designed to reinforce the concept of being connected. The reader who is uninterested in this aspect may desire to move ahead to the differential equation exercises.

Let

$$A = \{(x, y)|x^2 + y^2 \leq 1\}$$
$$B = \{(x, y)|(x - 1)^2 + y^2 \leq 1\}$$
$$C = \{(x, y)|x^2 + y^2 \leq \tfrac{1}{2}\}$$
$$A' = \{(x, y)|x^2 + y^2 < 1\}$$
$$B' = \{(x, y)|(x - 1)^2 + y^2 < 1\}$$
$$C' = \{(x, y)|x^2 + y^2 < \tfrac{1}{2}\}$$
$$R = \{(x, y)|x = 0\}.$$

1. Which of the following sets are connected?

   (a) $A \cup B$

   (b) $A \backslash C$ (Recall that $A \backslash C$ is the set of points of A that are not in C.)

   (c) $A \cup B'$ (d) $(B \cup C) \backslash R$

   (e) $B \cup C'$ (f) $A \backslash C'$

   (g) $A' \cup B'$ (h) $A \backslash B$

   (i) $(A \cup B) \backslash C$ (j) $A \backslash R$

   (k) $B \backslash R$ (l) $(A \backslash R) \cup B$

2. Which of the sets in Exercise 1 form regions?

3. Which of the sets in Exercise 1 are simply connected?

4. Obtain (7.2) from (7.1).

5. Give a formal proof that the omega limit set of a trajectory that is a periodic orbit is itself.

6. Show that if there are no periodic trajectories, the omega limit set of a bounded trajectory contains a critical point.

7. Show that if the omega limit set of a bounded trajectory contains an asymptotically stable critical point, then it consists of only that point.

8. Show that the system

   $$x' = y$$
   $$y' = 2(1 - xy)$$

   has no limit cycles.

9. Let $\Gamma(t) = (\varphi(t), \psi(t))$ be a trajectory for a solution of (7.3). Show that if P is an

omega limit point of Γ, then the trajectory of (7.3) with $(x(0), y(0)) = P$ is a subset of the omega limit set of Γ. (*Hint:* If $\Gamma(t_n) \to P$, consider $\Gamma(t_n + T)$.)

10. Use the Poincaré-Bendixson theorem to prove that the system

$$x' = x + y - x(x^2 + y^2)\cos^2(x^2 + y^2)$$
$$y' = -x + y - y(x^2 + y^2)\cos^2(x^2 + y^2)$$

has a limit cycle.

8. An Example: Lotka-Volterra Competition

The growth of a population with limited nutrients is frequently described by a logistic equation,

$$z' = \alpha z \left(1 - \frac{z}{K}\right). \tag{8.1}$$

The parameter α is called the *growth rate* and the parameter K, the *carrying capacity*. A solution of (8.1) with a positive initial condition $z(0) = z_0 > 0$ satisfies $\lim_{t\to\infty} z(t) = K$. For example, Figure 8.1 and 8.2 show two examples of biological data fit by a solution of an equation (8.1).

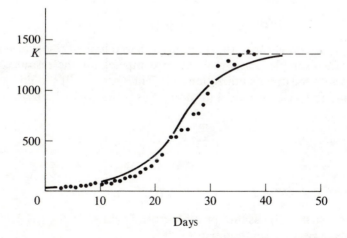

Figure 8.1 The growth of a population of *Hydra* in a mixed culture of freshwater organisms and a logistic curve fitted to the data. (Reproduced, by permission of the publisher, from Christiansen and Fenchel, *Theories of Populations in Biological Communities*. (Berlin: Springer-Verlag, 1977), page 3.)

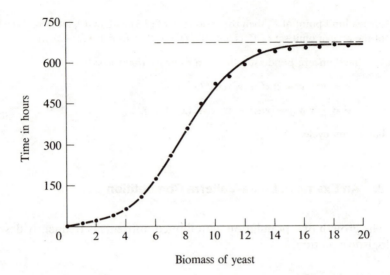

Figure 8.2 A comparison of the growth of yeast in culture with logistic growth. (Reproduced, by permission of the publisher, from J. Maynard Smith, *Models in Ecology.* Cambridge, MA: Cambridge University Press, 1974), page 18.)

Two populations that do not compete can be described by the uncoupled system (where $\beta_i = \alpha_i/K_i$)

$$x' = \beta_1 x(K_1 - x)$$
$$y' = \beta_2 y(K_2 - y). \tag{8.2}$$

Suppose the carrying capacity represents a shared limited resource—space or a nutrient, for example. Each population competes for the resource and each interferes with the other's utilization of it. The system (8.2) should be altered to reflect this, and a classic model for doing this is the *Lotka-Volterra competition equations,*

$$x' = \beta_1 x(K_1 - x - \lambda_1 y)$$
$$y' = \beta_2 y(K_2 - y - \lambda_2 x) \tag{8.3}$$
$$x(0) \geq 0, \ y(0) \geq 0.$$

The system (8.3) has four possible critical points, $(0, 0)$, $(0, K_2)$, $(K_1, 0)$, and (if it exists) a positive solution pair of

$$x + \lambda_1 y = K_1$$
$$\lambda_2 x + y = K_2. \tag{8.4}$$

The linearization corresponding to (8.3) takes the form

$$z' = Az$$

where A is the matrix

$$\begin{pmatrix} \beta_1(K_1 - 2\hat{x} - \lambda_1 \hat{y}) & -\beta_1 \lambda_1 \hat{x} \\ -\beta_2 \lambda_2 \hat{y} & \beta_2(K_2 - 2\hat{y} - \lambda_2 \hat{x}) \end{pmatrix} \tag{8.5}$$

and $(\hat{x}, \hat{y})$ is one of the aforementioned critical points. The local stability of the critical points is determined by the eigenvalues μ_i of the matrix above. For example, at $(\hat{x}, \hat{y}) = (0, 0)$, $\mu_i = \beta_i K_i > 0$, so the origin is strongly unstable—a "repeller." At $(\hat{x}, \hat{y}) = (0, K_2)$ and $\mu_1 = -\beta_2 K_2 < 0$, $\mu_2 = \beta_1(K_1 - \lambda_1 K_2)$, and $(\hat{x}, \hat{y})$ is either a saddle or a stable node depending on the sign of $K_1 - \lambda_1 K_2$. Similarly, at $(\hat{x}, \hat{y}) = (K_1, 0)$, $\mu_1 = -\beta_1 K_1 < 0$ and $\mu_2 = (K_2 - \lambda_2 K_1)\beta_2$, and again the critical point is either a saddle or a stable node, depending on the sign of $K_2 - \lambda_2 K_1$.

Solutions of (8.4) are (assuming $\lambda_1 \lambda_2 \neq 1$)

$$\hat{x} = \frac{K_1 - \lambda_1 K_2}{1 - \lambda_1 \lambda_2}$$

$$\hat{y} = \frac{K_2 - \lambda_2 K_1}{1 - \lambda_1 \lambda_2}. \tag{8.6}$$

This critical point has biological meaning only if $\hat{x} > 0$, $\hat{y} > 0$ (populations are inherently nonnegative). Four cases need to be considered; see Figure 8.3, corresponding to the four possible graphs of the lines given by (8.4) (these are called the *isoclines* of (8.3)).

If $K_1 > \lambda_1 K_2$ and $K_2 < \lambda_2 K_1$, then the interior critical point does not exist in the first quadrant. $(K_1, 0)$ is a stable node, while $(0, K_2)$ is a saddle. Since there is no interior critical point, there can be no closed trajectory interior to the positive quadrant (Theorem 7.3). Thus, every trajectory except one (with initial conditions of the form $(0, y_0)$, $y_0 > 0$) tend to the stable node, and thus the population x wins and y loses—$\lim_{t \to \infty} x(t) = K_1$ and $\lim_{t \to \infty} y(t) = 0$. Similarly, if $\lambda_1 K_2 > K_1$ and $\lambda_2 K_1 < K_2$, the interior critical point does not exist but the above outcome is reversed—x loses and y wins the competition.

Suppose now that the interior critical point given by (8.4) exists in the first quadrant (Figure 8.3c or 8.3d). The matrix of the linearization takes the form

$$A = \begin{pmatrix} -\beta_1 \hat{x} & -\beta_1 \lambda_1 \hat{x} \\ -\beta_2 \lambda_2 \hat{y} & -\beta_2 \hat{y} \end{pmatrix}$$

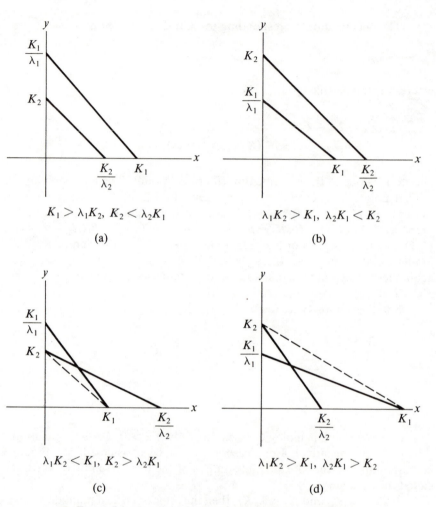

Figure 8.3 Isoclines for Lotka-Volterra competition.

with characteristic equation

$$\mu^2 + \mu(\beta_1 \hat{x} + \beta_2 \hat{y}) + \beta_1 \beta_2 \hat{x}\hat{y}(1 - \lambda_1 \lambda_2) = 0.$$

The roots are

$$\mu_{\pm} = \frac{-(\beta_1 \hat{x} + \beta_2 \hat{y}) \pm \sqrt{(\beta_1 \hat{x} + \beta_2 \hat{y})^2 - 4(1 - \lambda_1 \lambda_2)\beta_1 \beta_2 \hat{x}\hat{y}}}{2}.$$

If $\lambda_1 \lambda_2 > 1$, the roots are of opposite sign and $(\hat{x}, \hat{y})$ is a saddle point and hence unstable. Since we are assuming that the critical point is in the first quadrant, it follows that $K_1 < K_2 \lambda_1$ and $K_2 < K_1 \lambda_2$, or $(0, K_2)$ is stable and $(K_1, 0)$ is stable. Whether x or y wins the competition is a function of the initial conditions.

If $\lambda_1 \lambda_2 < 1$, both roots are negative and the interior critical point is a stable node. If $\lambda_1 \lambda_2 < 1$, then $K_1 > \lambda_1 K_2$ and $K_2 > \lambda_2 K_1$ and both $(K_1, 0)$ and $(0, K_2)$ are saddle points.

To make either of the preceding cases global, it is necessary to show that there are no limit cycles. Then, in the last case, it would follow from Theorem 7.1 that all solutions with initial conditions in the interior of the first quadrant tend to the (unique) interior critical points as t tends to infinity. Thus, the populations can coexist and tend to a constant "share" of the limited resource.

Since there is an interior critical point, a closed trajectory—a limit cycle—is a possibility. The Bendixson-Dulac criterion (Theorem 7.4) can be used to eliminate this possibility. Choose $\beta(x, y) = 1/xy$ and G to be the positive quadrant, $\{(x, y) | x > 0, y > 0\}$. Then

$$\frac{\partial}{\partial x} \left(\frac{1}{xy} \cdot \beta_1 x (K_1 - x - \lambda_1 y) \right) = \beta_1 \frac{\partial}{\partial x} \left[\frac{K_1}{y} - \frac{x}{y} - \lambda_1 \right]$$

$$= -\frac{\beta_1}{y}$$

and, similarly,

$$\frac{\partial}{\partial y} \left(\frac{1}{xy} \cdot \beta_2 y (K_2 - y - \lambda_2 x) \right) = -\frac{\beta_2}{x},$$

so that

$$\frac{\partial}{\partial x} (\beta(x, y) f(x, y)) + \frac{\partial}{\partial y} (\beta(x, y) g(x, y)) = -\frac{\beta_1}{y} - \frac{\beta_2}{x} < 0.$$

By Theorem 7.4, there can be no periodic solutions in the positive quadrant and the above analysis becomes a global one—a bounded solution must tend to the stable critical point(s). (It is not difficult to show that all solutions are eventually inside the rectangle $R = \{(x, y) | 0 \le x \le K_1, 0 \le y \le K_2\}$ and hence all solutions of (8.3) are bounded.)

Phase plane diagrams for the four cases are given in Figure 8.4.

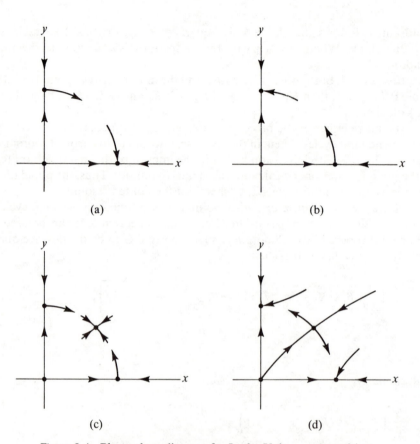

Figure 8.4 Phase plane diagram for Lotka-Volterra competition.

9. An Example: The Simple Pendulum

Consider a weight suspended on a rod attached to a pivot with a perfect (frictionless) bearing—a simple pendulum, as in Figure 9.1. The weight—sometimes called a bob—is constrained to move in a circular arc, as in Figure 9.2. If the angle with the vertical is denoted by θ, then Newton's second law ($F = ma$) dictates that the angle at time t is given as the solution of the initial value problem

$$mL\theta'' = mg \sin(\theta)$$

$$\theta(0) = \theta_0$$

$$\theta'(0) = \theta_1.$$

See Figure 9.3.

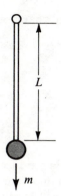

Figure 9.1 The simple pendulum.

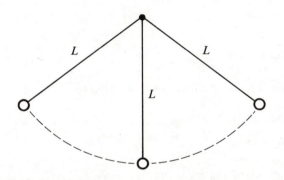

Figure 9.2 Constrained motion of the pendulum.

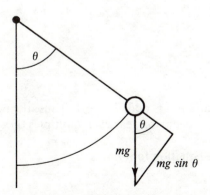

Figure 9.3 Resolution of forces for the simple pendulum.

Canceling mass from both sides and changing the time scale ($t = \sqrt{L/g}\tau$) eliminates the parameters, giving

$$\theta'' + \sin\theta = 0, \qquad ' = \frac{d}{d\tau} \tag{9.1}$$

$$\theta(0) = \theta_0 \tag{9.2}$$

$$\theta'(0) = \theta_1. \tag{9.3}$$

The approximation $\sin(\theta) = \theta$ would produce the equation used in Chapter 1 for the pendulum, but here we wish to illustrate our phase techniques on the nonlinear differential equation. Writing (9.1) as a system in the usual way yields

$$\begin{aligned} x' &= y \\ y' &= -\sin(x), \end{aligned} \tag{9.4}$$

$$\begin{aligned} x(0) &= x_0 = \theta_0 \\ y(0) &= y_0 = \theta_1. \end{aligned} \tag{9.5}$$

The critical points are determined as the solution of

$$y = 0$$
$$\sin(x) = 0$$

and hence are points of the form $(n\pi, 0)$ $n = 0, \pm 1, \pm 2, \ldots$. Analysis of the linearization about the critical points requires the eigenvalues of

$$\begin{bmatrix} 0 & 1 \\ -\cos(x) & 0 \end{bmatrix}_{\substack{x=n\pi \\ y=0}},$$

which are roots of

$$\lambda^2 + \cos(n\pi) = 0.$$

If n is even, $\lambda = \pm i$, and if n is odd, $\lambda = \pm 1$. Thus, the odd multiples of π are saddle points, while the behavior near even multiples of π is indeterminate and requires further analysis.

The phase plane equation for (9.4) is

$$\frac{dy}{dx} = \frac{-\sin(x)}{y}.$$

The variables separate, giving

$$\frac{y^2}{2} - \frac{y_0^2}{2} = \cos(x) - \cos(x_0).$$

The equation of a trajectory is then given by

$$y^2 = 2\cos(x) - 2\cos(x_0) + y_0^2.$$

If we place the initial conditions at $(0, y_0)$, then this becomes

$$y^2 = 2\cos(x) - 2 + y_0^2. \tag{9.6}$$

The curve corresponding to (9.6) reaches the x-axis at a point x such that

$$x = \cos^{-1}\left(\frac{2 - y_0^2}{2}\right).$$

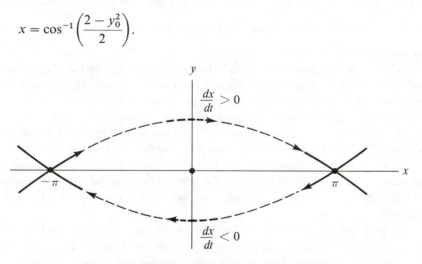

Figure 9.4 Saddle connections for the pendulum.

We restrict our analysis to $-\pi \le x \le \pi$; see Figure 9.4. Since $(-\pi, 0)$ and $(\pi, 0)$ are saddle points, we expect two trajectories entering and two leaving each point—it is natural to ask if they can be "connected." More precisely, is there a set of initial conditions (x_0, y_0) such that if $x(t)$, $y(t)$ is the solution through (x_0, y_0), $\lim_{t \to \infty} (x(t), y(t)) = (\pi, 0)$ and $\lim_{t \to -\infty} (x(t), y(t)) = (-\pi, 0)$? (Of course, we ask the same question with the critical points reversed.)

To see the existence of such a trajectory, take initial conditions $x_0 = 0$, $y_0 = \pm 2$. Then the loci of the two trajectories are given by

$$y^2 = 2\cos(x) + 2$$

and hence

$$\lim_{x \to \pm\pi} y(x) = 0.$$

Thus, we can fill in the dotted lines in Figure 9.4. These correspond to solutions such that

$$\lim_{t \to \pm\infty} \theta(t) = \pm\pi,$$

since no trajectory may reach a critical point in finite time. Physically, it means imparting exactly the correct amount of energy (the correct initial velocity in our case) so that the pendulum approaches the upright position as $t \to \infty$. Slightly more energy will send it "over the top"; slightly less will cause it to "fall back." That the arrows in (9.4) are correct may be seen from the fact that $\left.\dfrac{dx}{dt}\right|_{x=0}$ has the same sign as y.

What about trajectories "inside" the region bounded by the above trajectories? $(0,0)$ is a critical point, but what about the behavior of nearby solutions? Again, consider an initial condition of the form $(0, y_0)$. If $0 < y_0 < 2$, then $y(x) = 0$ will have two roots, one positive, one negative, occurring at $x_{\pm} = \cos^{-1}((2 - y_0^2)/2)$. Since $(x_+, 0)$ and $(x_-, 0)$ are not critical points, the trajectory continues through them. Furthermore, the trajectory with initial conditions $(0, -y_0)$ also passes through $(0, \cos^{-1}((y_0^2 - 2)/2)$ and, since trajectories cannot cross, these two trajectories must be the same. The trajectory then is a closed curve and corresponds to a periodic solution—the "usual" oscillations of a pendulum.

If a solution has initial conditions $(0, y_0)$, $|y_0| > 2$, then $y(x) = \sqrt{2\cos(x) - 2 + y_0^2}$ has no zeros—$y(x)$ remains positive or negative. These are the oscillations caused by giving an initial velocity (y_0) large enough to pass "over the top," after which the bob will continue to go over the top.

Since $\cos x$ is periodic, these trajectories repeat in each interval $(2n - 1)\pi$, $(2n + 1)\pi$, and

$$\left.\frac{dy}{dx}\right|_{x=n\pi} = 0.$$

These trajectories also correspond to periodic solutions, but they are solutions where the direction of motion is never reversed—solutions for which $\lim_{t \to \infty} \theta(t) = \pm\infty$.

The final phase plane result is shown in Figure 9.5.

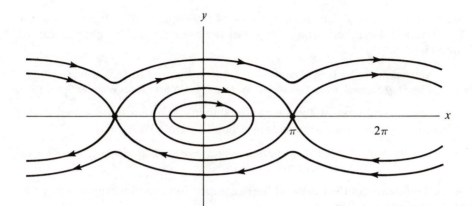

Figure 9.5   Phase plane diagram for the pendulum.

EXERCISES

1. Use Newton's second law to derive the equation for the motion of a pendulum. (*Hint:* Consult a physics book.)

2. Carry out the change-of-time-scale calculations needed to obtain equation (9.1).

3. Analyze the motion of the *damped* pendulum

$$L\theta'' = -g\sin(\theta) - k\theta', \quad k > 0.$$

Draw a phase plane diagram.

4. Give an argument that a nontrivial trajectory may not reach a critical point in finite time. (*Hint:* Use the uniqueness theorem.)

REFERENCES

The classification of critical points follows that in Chapter 15 of

Coddington, E. A., and N. Levinson. *Theory of Ordinary Differential Equations.* New York: McGraw-Hill, 1955.

See also Chapter 3 of

Hale, J. K. Ordinary Differential Equations. New York: Wiley-Interscience, 1969.

The qualitative approach is emphasized in the intermediate-level textbook

Hirsch, M. W., and S. Smale. *Differential Equations, Dynamical Systems, and Linear Algebra.* New York: Academic Press, 1974.

Much of the material (including the idea of same qualitative structure and the Bendixson-Dulac criterion) and many of the examples in this chapter can be found in

Andronov, A. A., E. A. Leontovich, I. I. Gordon, and A. G. Maier. *Qualitative Theory of Second Order Dynamical Systems*. New York: John Wiley & Sons, 1973.

A complete treatment of the Liapunov stability theorem is presented in the following. (See also the aforementioned book by Hale.)

LaSalle, J. P. *The Stability of Dynamical Systems*. Philadelphia: Society for Industrial and Applied Mathematics, 1976.

A good reference for the Lotka-Volterra competition equations (and many other ecological problems) is

Hutchinson, G. E. *An Introduction to Population Ecology*. New Haven, CT: Yale University Press, 1978.

The reader interested in the relationship between mathematics and biology may be interested in

Freedman, H. I. *Deterministic Mathematical Models in Population Ecology*. New York: Marcel Dekker, Inc., 1980.

Hoppensteadt, F. C. *Mathematical Models of Population Biology*. Cambridge, MA: Cambridge University Press, 1982.

Waltman, P. *Competition Models in Population Biology*. Philadelphia: Society for Industrial and Applied Mathematics, 1983.

An analysis of the pendulum appears in Chapter 1 of

Haberman, R. *Mathematical Models*. Englewood Cliffs, NJ: Prentice-Hall, 1977.

3

Existence Theory

1. Introduction

A major theoretical question in the study of ordinary differential equations is "when do solutions exist?" In this chapter, we study this question and the closely related ones of the uniqueness of solutions and the variation of solutions with their initial conditions. The basic theorem can be understood in terms of the simplest differential equation

$$y'(t) = f(t), \qquad ' = \frac{d}{dt} \tag{1.1}$$

where $f(t)$ is a continuous function defined on the real line, R. The fundamental theorem of calculus allows us to conclude that *any* antiderivative of $f(t)$ will be a solution of (1.1). Thus there is an entire family of functions, differing from one another by a constant, that when substituted for $y(t)$ in (1.1) produce an identity. This family can be written

$$y(t) = y_0 + \int_{t_0}^{t} f(s)\, ds \tag{1.2}$$

where t_0 and y_0 are constants. The pair (t_0, y_0) specifies a point in the plane and (1.2), for fixed (t_0, y_0), specifies the antiderivative of $f(t)$ that passes through that point. The fundamental theorem of calculus allows us to conclude that this is the only antiderivative with this property. The model for the basic theorem is

---

THEOREM 1.1

> **If $f(t)$ is a continuous function, then there is exactly one differentiable function that satisfies (1.1) and the initial condition $y(t_0) = y_0$.**

Of course, existence has been proved by exhibiting the solution, which is too much to expect in the general case. For example, consider the initial value problem

$$y' = f(t, y)$$
$$y(t_0) = y_0.$$

(1.3)

We cannot exhibit a solution, but we can hope to prove a *theorem* that asserts the existence of a unique solution if the function $f(t, y)$ is not too badly behaved. Such a theorem is the main goal of this chapter. We can also vary the "initial time," t_0, and the "initial point," y_0, and inquire how the solution changes as these constants change. The constants are often measured quantities and the meaning of physical or biological experiments will depend on the fact that the solution will vary "continuously" as these quantities change. A precise statement of this fact will also be expressed as a general theorem about differential equations.

In Section 2 the general initial value problem is changed to a problem that is easier to handle, an equivalent integral equation. A preliminary discussion of the appropriate conditions for the right-hand side of (1.3) also takes place and a new concept, that of a metric space, is introduced. In Section 3, a very powerful abstract theorem, the contraction mapping theorem, is proved. Sections 4 and 5 apply this theorem to establish existence and uniqueness theorems for a wide class of initial value problems. Section 6 treats a new type of problem, a boundary value problem, with the same technique.

2. Preliminaries

Consider the initial value problem

$$y' = f(t, y)$$

(2.1)

$$y(t_0) = y_0 \tag{2.2}$$

where $f(t, y)$ is continuous on a closed rectangle in R^2, which we denote by Ω : $|t - t_0| \leq a$, $|y - y_0| \leq b$. Our goal is to prove that there exists a solution of (2.1), (2.2) on an interval I including t_0 whose graph lies in Ω. It turns out to be more convenient to work with an integral equation rather than with the initial value problem (2.1), (2.2).

A continuous function $\varphi(t)$ defined on an interval I is said to be a solution on I of the integral equation

$$y = y_0 + \int_{t_0}^{t} f(s, y)\, ds \tag{2.3}$$

if for every $t \in I$, $f(t, \varphi(t))$ is defined, and

$$\varphi(t) = y_0 + \int_{t_0}^{t} f(s, \varphi(s))\, ds.$$

THEOREM 2.1

Let $f(t, y)$ be continuous. A function $\varphi(t)$, defined on an interval I, is a solution of (2.1), (2.2) on I if and only if it is a continuous solution of (2.3) there.

Proof: Let $\varphi(t)$ be a solution of (2.1), (2.2). Let $K(t) = f(t, \varphi(t))$. $K(t)$ is continuous, and since

$$\varphi'(t) = K(t),$$

integrating from t_0 to t gives

$$\int_{t_0}^{t} \varphi'(s)\, ds = \int_{t_0}^{t} K(s)\, ds$$

or

$$\varphi(t) = \varphi(t_0) + \int_{t_0}^{t} K(s)\, ds$$

$$= y_0 + \int_{t_0}^{t} f(s, \varphi(s))\, ds.$$

Thus $\varphi(t)$ is a solution of (2.3) and is continuous because it is differentiable. Conversely, if $\varphi(t)$ is a solution of (2.3), that is, if

$$\varphi(t) = y_0 + \int_{t_0}^{t} f(s, \varphi(s))\, ds,$$

then $\varphi(t_0) = y_0$, so (2.2) is satisfied. Further, $f(t, \varphi(t))$ is a continuous function of t, since f and φ are continuous, so $\varphi(t)$ is differentiable and (by the fundamental theorem of calculus)

$$\varphi'(t) = f(t, \varphi(t)).$$

Therefore, $\varphi(t)$ satisfies (2.1) and the proof is complete.

Unfortunately, the continuity of $f(t, y)$ is not a sufficiently strong condition to guarantee that (2.1), (2.2) has a unique solution. To ensure this we must add a condition on the function $f(t, y)$—the Lipschitz condition. A function $f(t, y)$ is said to satisfy a *Lipschitz condition* in y in a region Ω if there is a constant K such that for every pair of points (t, y_1) and (t, y_2) in Ω,

$$|f(t, y_1) - f(t, y_2)| \leq K|y_1 - y_2|.$$

First, if $f(t, y) = p(t)y + q(t)$, $p(t)$ continuous on a closed interval I, then $f(t, y)$ satisfies a Lipschitz condition. To see this, let (t, y_1), (t, y_2), $t \in I$, be a pair of points. Then,

$$|f(t, y_1) - f(t, y_2)| = |p(t)y_1 + q(t) - p(t)y_2 - q(t)|$$
$$= |p(t)(y_1 - y_2)|$$
$$= |p(t)||y_1 - y_2|.$$

Since $p(t)$ is continuous on a closed interval I, it is bounded there. Let K be a number such that $|p(t)| \leq K$ on I. Using this bound in the equality above produces

$$|f(t, y_1) - f(t, y_2)| \leq K|y_1 - y_2|,$$

and so $f(t, y)$ satisfies a Lipschitz condition.

For another example, suppose that $f(t, y)$ is such that $\partial f/\partial y$ exists and, moreover, there exists a constant K such that for all (t, y) in a region Ω, $|(\partial f/\partial y)(t, y)| \leq K$. This is true, for example, if $\partial f/\partial y$ is continuous and Ω is a closed rectangle. Then for (t, y_1), (t, y_2) in Ω, the mean value theorem asserts that

$$|f(t, y_1) - f(t, y_2)| = \left| \frac{\partial f}{\partial y}(t, \eta) \right| |y_1 - y_2|$$

for some $(t, \eta) \in \Omega$. Since the partial derivative is bounded for $(t, y) \in \Omega$, it follows that

$$|f(t, y_1) - f(t, y_2)| \le K|y_1 - y_2|$$

and f satisfies a Lipschitz condition in Ω. Differentiability is not necessary for a Lipschitz condition to hold ($f(t, y) = |y|$ is not differentiable at $y = 0$ but it does satisfy such a condition), but boundedness of the partial derivative is a useful, sufficient condition for $f(t, y)$ to satisfy a Lipschitz condition. This Lipschitz condition will be a basic hypothesis in the statement of the existence and uniqueness theorem.

The abstract theorem proved in the next section requires the environment of a "metric space." This abstract idea is valuable in many areas of mathematics, so we pause in our study of differential equations to pursue it. Its power is that it allows us to think of functions as points in a set, just as we think of ordered pairs of real numbers as points in a plane. The concept of distance, or metric, is needed to allow the definitions of analysis (for example, the definition of limit) to be formulated. The basic ingredients are a set and a notion of distance. We proceed to make this precise.

Let M be a set. A function ρ is called a *metric* (or a *distance*) if it associates with each pair of elements (x, y) of M a real number $\rho(x, y)$ with the following properties:

1. $\rho(x, y) \ge 0$ and $\rho(x, y) = 0$ if and only if $x = y$;

2. $\rho(x, y) = \rho(y, x)$; and

3. $\rho(x, y) \le \rho(x, z) + \rho(y, z)$ for every x, y, z in M.

The pair (M, ρ) is said to be a *metric space*. When the metric is understood we sometimes simply say that M is a metric space. For example, the set of real numbers with $\rho(x, y) = |x - y|$ is a metric space, and the set of complex numbers with $\rho(z, w) = |z - w|$ (the modulus of the difference) is also a metric space. The set of constant $n \times n$ matrices with $\rho(A, B) = \|A - B\|$ (recall the definition of norm of a matrix given in Chapter 1) is a metric space. The same set of elements can be given different metrics and hence can form different metric spaces. Consider R^2 the set of points in the plane and let $P_\alpha = (x_\alpha, y_\alpha)$. We can define

$$\rho(P_1, P_2) = \sqrt{(x_1 - x_2)^2 + (y_1 - y_2)^2},$$

the usual Euclidean, or straight-line, distance (which is the same metric as given to the complex numbers). We can also define

$$\rho(P_1, P_2) = |x_1 - x_2| + |y_1 - y_2|.$$

This is called the "taxicab" metric—it is the distance traveled on a city street between two points (given perfect city streets). This metric is frequently convenient in problems in differential equations.

A more interesting metric space is the following. Let M be the set of all continuous functions on an interval $[a, b]$ and define $\rho(x, y) = \max_{a \le t \le b} |x(t) - y(t)|$. The quantity $|x(t) - y(t)|$ is a continuous function and hence has a maximum on a closed interval. We show that this function satisfies the conditions above. Clearly, $|x(t) - y(t)| \ge 0$ for every $t \in [a, b]$, so the maximum is also ≥ 0. If $x(t) = y(t)$, then $\rho(x, y) = 0$, and if $\rho(x, y) = 0$, that is, $\max_{t \in [a, b]} |x(t) - y(t)| = 0$, then $x(t)$ is not different from $y(t)$ at any $t \in [a, b]$, so $x(t)$ and $y(t)$ are the same function. Further, $|x(t) - y(t)| = |y(t) - x(t)|$, so $\rho(x, y) = \rho(y, x)$. To see property (3), called the *triangle inequality*, observe that for any functions $x(t)$, $y(t)$, and $z(t)$,

$$|x(t) - y(t)| = |x(t) - z(t) + z(t) - y(t)|$$
$$\le |x(t) - z(t)| + |z(t) - y(t)|,$$

so that

$$\rho(x, y) = \max_{t \in I} |x(t) - y(t)|$$
$$\le \max_{t \in I} \{|x(t) - z(t)| + |y(t) - z(t)|\}$$
$$\le \max_{t \in I} |x(t) - z(t)| + \max_{t \in I} |y(t) - z(t)|$$
$$= \rho(x, y) + \rho(y, z).$$

This space, which is customarily denoted by $C[a, b]$, and some similar spaces, are frequently used in the study of differential equations. Note that $x \equiv 0$, the zero function, is in $C[a, b]$ and that $\rho(y, 0) = \max_{t \in [a, b]} |y(t) - 0| = \max_{t \in [a, b]} |y(t)|$.

The metric allows us to define the concepts of convergence and of limit. A sequence $\{x_n\}$ of points in a metric space M is said to be *convergent* (also called a *Cauchy sequence*) if for every $\varepsilon > 0$ there exists a positive integer N such that, for $n \ge N$ and $m \ge N$, $\rho(x_n, x_m) < \varepsilon$. An element of M is said to be the *limit* of a convergent sequence $\{x_n\}$ if for every $\varepsilon > 0$ there is a positive integer N such that $\rho(y, x_n) < \varepsilon$ for all $n \ge N$, that is, if there is a $y \in M$ such that the sequence of real numbers $\{\rho(y, x_n)\}$ converges to zero. We write $\lim_{n \to \infty} x_n = y$. Although when dealing with real numbers a convergent sequence always converges "to" something, that is, a convergent sequence has a limit, this is not true in metric spaces in general. For example, the rational numbers in $[0, 1]$ form a metric space with $\rho(x, y) = |x - y|$, but a sequence of rational numbers can converge to an irrational number. Such a sequence is convergent (is a Cauchy sequence) but it does not converge "to" something in the space. Fortunately, the space $C[a, b]$

has the property that convergent sequences have a limit. A metric space M is said to be *complete* if every convergent sequence has a limit in the space.

THEOREM 2.2

The space $C[a, b]$ is complete.

The proof of this theorem is quite technical. The reader may wish to accept the fact that $C[a, b]$ is complete and omit the proof.

Proof: Let $\{x_n\}$ be a convergent sequence in $C[a, b]$; that is, each $x_n = x_n(t)$ is a continuous function on $[a, b]$ and for $\varepsilon > 0$ there is an N such that $\max_{t \in [a,b]} |x_n(t) - x_m(t)| = \rho(x_n, x_m) < \varepsilon$ for $n, m \geq N$. We must show that there exists a function, $y(t)$, continuous on $[a, b]$, such that $\lim_{n \to \infty} x_n = y$.

For a fixed t, $\{x_n(t)\}$ is a sequence of real numbers, and since

$$|x_n(t) - x_m(t)| \leq \max_{t \in [a,b]} |x_n(t) - x_m(t)| = \rho(x_n, x_m),$$

$\{x_n(t)\}$ is a convergent sequence of real numbers for every choice of t. For each $t \in [a, b]$, define

$$y(t) = \lim_{n \to \infty} x_n(t).$$

To prove the theorem we need only show that $y(t)$ is continuous on $[a, b]$, that is, that $y(t)$ is in the space $C[a, b]$, and that x_n converges to y in the metric (the convergence above is "pointwise" convergence).

We show first a consequence of the convergence. Let $\varepsilon > 0$ and let N be the number given in the definition of convergence. Then for any choice of $n, m \geq N$, and any t in $[a, b]$,

$$|x_n(t) - x_m(t)| \leq \rho(x_n, x_m) < \varepsilon.$$

We fix n and take the limit on the left-hand side as $m \to \infty$; that is,

$$\lim_{m \to \infty} |x_n(t) - x_m(t)| \leq \varepsilon$$

or

$$|x_n(t) - y(t)| \leq \varepsilon,$$

$n \geq N$, for all t in $[a, b]$. Note that if $y(t)$ were an element of $C[a, b]$, that is, if $y(t)$ were continuous, this would be sufficient to conclude that $\rho(x_n, y) \to 0$ or that x_n converges to y in the metric space $C[a, b]$.

We now establish the continuity of $y(t)$ at each point of $[a, b]$. Let $\varepsilon > 0$ and $t_0 \in [a, b]$ be fixed. Corresponding to $\varepsilon/3$ there is an N such that $|x_n(t) - y(t)| \leq \varepsilon/3$ for all $t \in I$ and for $n \geq N$. Fix $n > N$. Since $x_n(t)$ is a continuous function, corresponding to $\varepsilon/3$, there is a $\delta > 0$ such that if $|t - t_0| < \delta$, then $|x_n(t) - x_n(t_0)| < \varepsilon/3$. Hence it follows that

$$|y(t) - y(t_0)| \leq |y(t) - x_n(t)| + |x_n(t) - x_n(t_0)| + |x_n(t_0) - y(t_0)|.$$

If $|t - t_0| < \delta$, this yields

$$|y(t) - y(t_0)| < \frac{\varepsilon}{3} + \frac{\varepsilon}{3} + \frac{\varepsilon}{3} = \varepsilon.$$

Thus $y(t)$ is a continuous function, and therefore is an element of $C[a, b]$. As noted before, if $\varepsilon > 0$ there is an N such that if $n \geq N$,

$$|y(t) - x_n(t)| \leq \frac{\varepsilon}{3} < \varepsilon$$

for every t in $[a, b]$, and since $y(t)$ is continuous we can take the maximum over $[a, b]$ and obtain

$$\rho(y, x_n) \leq \frac{\varepsilon}{3} < \varepsilon.$$

Thus y is the limit of the sequence $\{x_n\}$ in the space $C[a, b]$.

Sometimes it is convenient to work on a subset of $C[a, b]$, the set B of continuous functions on $[a, b]$ that also satisfy $\rho(x, x_0) \leq \alpha$ for some element x_0 and some number α. (B is a set of bounded continuous functions.)

LEMMA 2.3

The space B is complete.

Proof: Let $\{x_n\}$ be a convergent sequence with $x_n \in B$. Since the x_n all belong to $C[a, b]$, x_n converges to $y \in C[a, b]$ by Theorem 3.1, and we need only show that $y \in B$ to prove the lemma.

Let $\varepsilon > 0$. Then there is an N such that if $n \geq N$, $\rho(x_n, y) < \varepsilon$. Hence

$$\rho(y, x_0) \leq \rho(y, x_n) + \rho(x_n, x_0) < \varepsilon + \alpha.$$

If for every $\varepsilon > 0$, $\rho(y, x_0) < \alpha + \varepsilon$, then $\rho(y, x_0) \leq \alpha$. (If there were a point t_1 such that $|y(t_1) - x_0(t_1)| > \alpha$, then we would choose $\varepsilon < |y(t_1) - x_0(t_1)| - \alpha$ and obtain a contradiction.) Thus, $y \in B$.

Finally, we note that basic properties like continuity can be defined for mappings on a metric space. Let $T: M \to M$. T is continuous at a point x if, for every $\varepsilon > 0$, there is a $\delta > 0$ such that if $\rho(x, y) < \delta$, then $\rho(Tx, Ty) < \varepsilon$. This reads like the definition given in calculus, except that metric has been substituted for absolute value.

EXERCISES

1. Find a Lipschitz constant for the following functions in appropriate domains.

   (a) $\sin(y)$

   (b) $\dfrac{1}{1 + y^2}$

   (c) $e^{-|y|}$

   (d) $\tan^{-1}(y)$

2. Show that $|x(t) - y(t)|$ is a continuous function if $x(t)$ and $y(t)$ are continuous functions.

3. Let M be the set of functions defined on $[0, 1]$ that have a continuous derivative there (one-sided derivatives at the endpoints). Define
   $$\rho(x, y) = \max_{t \in [0, 1]} \{|x(t) - y(t)| + |x'(t) - y'(t)|\}.$$
   Show that this is a metric space.

4. Show that, with the "taxicab" metric, ordered pairs of real numbers (points in the plane) form a metric space.

   *A sequence of functions $y_n(t)$ defined on $[0, 1]$ is said to **converge uniformly** if for $\varepsilon > 0$ there is an $N > 0$ such that if $n, m > N$, then $|y_n(t) - y_m(t)| < \varepsilon$ for every $t \in [0, 1]$.*

5. Show that the limit of a uniformly convergent sequence of continuous functions defined on $[0, 1]$ converges to a continuous function. (*Hint:* Follow the proof of Theorem 2.2.)

6. Show that uniform convergence of a sequence of continuous functions defined on $[0, 1]$ and convergence of the corresponding sequence in $C[0, 1]$ are equivalent concepts.

7. Let M be as in Exercise 3 and let $\rho(x, y) = \max_{[0, 1]} |x'(t) - y'(t)|$. Show that (M, ρ) fails to be a metric space. Let $\rho(x, y) = |x(0) - y(0)| + \max_{[0, 1]} |x'(t) - y'(t)|$. Is (M, ρ) now a metric space?

8. Let M be the set of continuous functions on $[0, 1]$ and define $\rho(x, y) = \int_0^1 |x(t) - y(t)| \, dt$. Does this define a metric space?

3. The Contraction Mapping Theorem

The contraction mapping theorem concerns operators that map a given metric space into itself. Given an element of the metric space M, an operator merely associates with it a unique element of M. Since the metric spaces of interest here are spaces of functions, we give an example of an operator on the space $C[a, b]$. For each $x \in C[a, b]$, define Tx to be the function

$$Tx(t) = \int_a^t x(s)\, ds, \qquad t \in [a, b].$$

Clearly, the image function, $y(t)$,

$$y(t) = Tx(t)$$

is a continuous function (in fact, a differentiable one), so $y \in C[a, b]$. Thus we write

$$T : C[a, b] \to C[a, b].$$

Note that not every element of $C[a, b]$ is the image of a point—T maps $C[a, b]$ into a subset of $C[a, b]$.

Let T be a mapping (an operator) from a metric space M into itself, $T : M \to M$. T is said to be a *contraction* if for each x, y in M,

$$\rho(Tx, Ty) \le \alpha \rho(x, y), \qquad 0 \le \alpha < 1.$$

Recall that T is continuous at x if for each $\varepsilon > 0$ there is a $\delta > 0$ such that $\rho(Tx, Ty) < \varepsilon$ if $\rho(x, y) < \delta$. Being a contraction mapping is a stronger property than being a continuous one. Let $T : M \to M$ be a contraction mapping. If $\varepsilon > 0$ is given, choose $\delta = \varepsilon$. Then, if $\rho(x, y) < \delta$,

$$\rho(Tx, Ty) \le \alpha \rho(x, y) < \alpha \delta < \varepsilon,$$

since $\alpha < 1$. Continuity can also be expressed in a sequential form. If

$$\lim_{n \to \infty} x_n = x,$$

then

$$\lim_{n \to \infty} Tx_n = Tx,$$

or, as we shall use it,

$$T\left(\lim_{n \to \infty} x_n\right) = \lim_{n \to \infty} Tx_n.$$

The proof of the equivalence of the two definitions is left as an exercise.
We now prove a theorem that is very useful in applied mathematics.

THEOREM 3.1 *(contraction mapping theorem)*
A contraction mapping T, defined on a complete metric space, has a unique fixed point; that is, there is a unique element, $x \in M$, such that $Tx = x$.

Proof: Let y_0 be any element of M. Let $y_n = Ty_{n-1}$, $n = 1, 2, \ldots$; that is, $y_1 = Ty_0$, $y_2 = Ty_1$, etc. We show that $\{y_n\}$ is convergent. Let m be an arbitrary positive integer. Then

$$\rho(y_{m+1}, y_m) = \rho(Ty_m, Ty_{m-1}) \leq \alpha\rho(y_m, y_{m-1})$$

$$= \alpha\rho(Ty_{m-1}, Ty_{m-2}) \leq \alpha^2\rho(y_{m-2}, y_{m-3}) \leq \cdots$$

$$\leq \alpha^m\rho(y_1, y_0).$$

Hence, for $k > 0$, repeated application of property (3) of a metric (the *triangle inequality*) gives

$$\rho(y_{n+k}, y_n) \leq \rho(y_{n+k}, y_{n+k-1}) + \rho(y_{n+k-1}, y_{n+k-2})$$

$$+ \cdots + \rho(y_{n+1}, y_n)$$

$$\leq (\alpha^{n+k-1} + \alpha^{n+k-2} + \cdots + \alpha^n)\rho(y_1, y_0).$$

Now since $0 < \alpha < 1$, the geometric series $1 + \alpha + \alpha^2 + \alpha^3 + \cdots + \alpha^n + \cdots$ is convergent and has sum $1/(1 - \alpha)$; that is,

$$\sum_{n=0}^{\infty} \alpha^n = \frac{1}{1 - \alpha}.$$

Thus

$$\alpha^n + \alpha^{n+1} + \cdots + \alpha^{n+k-1} = \alpha^n\left(\sum_{i=0}^{k-1} \alpha^i\right) \leq \frac{\alpha^n}{1 - \alpha}.$$

This yields that

$$\rho(y_{n+k}, y_n) \leq \frac{\alpha^n}{1 - \alpha}\rho(y_1, y_0) \tag{3.1}$$

for all $k > 0$. Since $0 < \alpha < 1$, $(\alpha^n/(1 - \alpha))\rho(y_1, y_2) \to 0$ as $n \to \infty$, so $\{y_n\}$ is convergent. Since M is complete, $\{y_n\}$ has a limit. Let y be the element of M such that $\lim_{n \to \infty} y_n = y$. We show that $y = Ty$, or that y is a fixed point. Note that

$$Ty = T\left(\lim_{n \to \infty} y_n\right)$$

$$= \lim_{n \to \infty} Ty_n,$$

by continuity of T. Also

$$\lim_{n \to \infty} Ty_n = \lim_{n \to \infty} y_{n+1} = y,$$

or,

$$Ty = y,$$

which completes the proof of existence.

Suppose there are two different points x and y with $x = Tx$, $y = Ty$. Then

$$\rho(x, y) = \rho(Tx, Ty) \leq \alpha\rho(x, y) < \rho(x, y)$$

or

$$\rho(x, y) < \rho(x, y),$$

which is a contradiction. Thus the fixed point is unique and the theorem is established.

Actually, some of the properties of the iterates developed in the proof are useful also. The following corollary gives an estimate of (an upper bound for) the distance of the nth iterate from the fixed point.

COROLLARY 3.2

The sequence of iterates converges for any starting point y_0. Further,

$$\rho(y_n, y) \leq \frac{\alpha^n}{1 - \alpha}\rho(y_1, y_0).$$

The last statement in the corollary, which follows from (3.1) (by letting $k \to \infty$), gives an error bound on the difference between the fixed point and the nth iterate in terms of α and the distance between the starting function and the first iterate.

EXERCISES

1. Show that the two definitions of continuity in this section are the same.

2. Define $T: C[0, 1] \to C[0, 1]$ by $(Tx)(t) = 1 + \int_0^t x(s)\, ds$. Is T a contraction?

3. Change the space in Exercise 2 to $C[0, \frac{1}{2}]$. Is T a contraction? Can you identify a function that is a fixed point?

4. Consider the operator in $C[0, 1]$,

$$Ty(t) = \int_0^t (t - s)y(s)\, ds.$$

Show that T is a contraction.

5. Let y_0 be the constant function 1 on the interval $[0, 1]$; that is, $y_0(t) = 1, t \in [0, 1]$. Use the operator in Exercise 4 and define

$$y_1 = Ty_0$$

$$y_2 = Ty_1$$

$$\vdots$$

$$y_n = Ty_{n-1}.$$

Construct $y_1(t), y_2(t), y_3(t), y_4(t)$. Can you conjecture the limit of $y_n(t)$ as $n \to \infty$?

6. Let M be the set of continuous functions on $[0, 1]$ and let

$$\rho(x, y) = \max_{t \in [0, 1]} \frac{|x(t) - y(t)|}{1 + t}.$$

Show that this defines a metric space.

7. Reconsider Exercise 2 with this new space (Exercise 6). This shows that changing the metric of the space can make an operator into a contraction.

4. The Initial Value Problem for One Scalar Differential Equation

It is now easy to prove a local existence theorem using the contraction mapping theorem.

THEOREM 4.1

Let $f(t, y)$ be continuous and satisfy a Lipschitz condition with Lipschitz constant K on $\Omega = \{(t, y) | |t - t_0| \leq a, |y - y_0| \leq b\}$ and let M be a number such that $|f(t, y)| \leq M$ for $(t, y) \in \Omega$. Choose $0 < \alpha < \min[1/K, b/M, a]$.

Then there exists a unique solution of

$$y' = f(t, y) \tag{4.1}$$

$$y(t_0) = y_0 \tag{4.2}$$

valid on $|t - t_0| \leq \alpha.$

Proof: We take the basic space to be $B = \{\varphi \,|\, \varphi \in C\,[t_0 - \alpha, \,t_0 + \alpha], \,\rho(\varphi, y_0) \leq b\}$, which was shown to be a complete metric space (Lemma 2.3). Define T as

$$T[\varphi](t) = y_0 + \int_{t_0}^t f(s, \varphi(s))\, ds;$$

that is, given $\varphi \in B$, the image of φ is the function (of t)

$$y_0 + \int_{t_0}^t f(s, \varphi(s))\, ds.$$

Since $f(t, \varphi(t))$ is defined for $t \in [t_0 - \alpha, \,t_0 + \alpha]$ and $\varphi \in B$ and is a continuous function of t and $\varphi(t)$, the image is continuous. Further,

$$|(T\varphi)(t) - y_0| \leq \left| \int_{t_0}^t |f(s, \varphi(s))|\, ds \right|$$

$$\leq M|t - t_0| \leq M\alpha < b.$$

Since $|(T\varphi)(t) - y_0|$ is continuous, it takes on its maximum on $[t_0 - \alpha, \,t_0 + \alpha]$, which is then $\leq b$; that is, $\rho(T\varphi, y_0) \leq b$, so $T\varphi \in B$ if $\varphi \in B$; that is, T maps B into B.

We show that T is a contraction mapping:

$$|(T\varphi - T\psi)(t)| = \left| y_0 + \int_{t_0}^t f(s, \varphi(s))\, ds - y_0 - \int_{t_0}^t f(s, \psi(s))\, ds \right|$$

$$\leq \left| \int_{t_0}^t |f(s, \varphi(s)) - f(s, \psi(s))|\, ds \right|$$

$$\leq K \left| \int_{t_0}^t |\varphi(s) - \psi(s)|\, ds \right|,$$

by the Lipschitz condition. Now

$$|\varphi(t) - \psi(t)| \leq \max_{t_0 - \alpha \leq t \leq t_0 + \alpha} |\varphi(t) - \psi(t)| = \rho(\varphi, \psi).$$

So

$$|T\varphi(t) - T\psi(t)| \leq K\rho(\varphi, \psi) \left| \int_{t_0}^t ds \right| = K\rho(\varphi, \psi)|t - t_0|$$

$$\leq K\alpha\rho(\varphi, \psi),$$

since $|t - t_0| \leq \alpha$, by hypothesis. If $|T\varphi(t) - T\psi(t)| \leq K\alpha\rho(\varphi, \psi)$ for every

t in $[t_0 - \alpha, t_0 + \alpha]$ then, since the maximum of the continuous function $(T\varphi - T\psi)(t)$ occurs for t in $[t_0 - \alpha, t_0 + \alpha]$,

$$\max_{t_0 - \alpha \le t \le t_0 + \alpha} |T\varphi(t) - T\psi(t)| \le K\alpha\rho(\varphi, \psi)$$

or

$$\rho(T\varphi, T\psi) \le K\alpha\rho(\varphi, \psi).$$

Since $K\alpha < 1$ by the choice of α, T is a contraction mapping and by Theorem 3.1 has a fixed point.

This fixed point is the unique solution of the integral equation

$$\psi(t) = y_0 + \int_{t_0}^{t} f(s, \psi(s)) \, ds$$

and hence is the unique solution of the initial value problem (4.1), (4.2) by Theorem 2.1.

The existence theorem proved here is of a local character in that the existence of a solution was established in a suitably small interval $I = [t_0 - \alpha, t_0 + \alpha]$. We could, of course, take this solution and examine new initial value problems at the points $t_0 - \alpha$ and $t_0 + \alpha$. The existence theorem then guarantees a solution in a suitably small interval about $t_0 - \alpha$ and about $t_0 + \alpha$, so the solution can be "extended." The process could then be repeated. However, it may turn out that the "suitably small" intervals get smaller and smaller, and that there is a point t_1 that cannot be reached. That this can indeed occur can be shown by considering the initial value problem

$$y' = 1 + y^2$$
$$y(0) = 0$$

whose solution is $y = \tan t$. The solution cannot be extended to an interval larger than $(-\pi/2, \pi/2)$ even though $1 + y^2$ is a "well-behaved" function.

Rather than having to go through the extension procedure for each function, it is convenient to have a theorem that provides sufficient conditions for a solution to exist for all t. Using the contraction mapping theorem, a global existence and uniqueness theorem is easily established under a stronger hypothesis.

THEOREM 4.2

Let $f(t, y)$ be continuous and satisfy a Lipschitz condition with constant K for every t and y. For real numbers t_0 and y_0, there exists a unique solution of the initial value problem (4.1), (4.2), valid for all t.

A function that satisfies the Lipschitz condition for all t and y *with the same Lipschitz constant* is said to be *uniformly Lipschitzian*. The function $f(t, y) = \sin y$ is such a function. This is in contrast to the function $f(t, y) = 1 + y^2$, used in the example above, which satisfies a Lipschitz condition in any closed and bounded region, but in which the Lipschitz constant changes as the region changes. Such functions are said to be *locally Lipschitzian*.

Proof of Theorem 4.2: Given t_0, y_0, Theorem 4.1 guarantees the existence of a solution $y(t)$, valid on $[t_0 - \alpha, t_0 + \alpha]$ with $y(t_0) = y_0$. We apply the extension procedure outlined in the first paragraph of this section. If that procedure does not reach every $t \in R$, there is a point t_1 such that no solution can be continued to $[t_0, t_1]$ or $[t_1, t_0]$. We assume that $t_1 > t_0$ and prove the existence of a unique solution of (4.1), (4.2) valid on the interval $[t_0, t_1]$. This contradiction shows that such a barrier as t_1 does not exist and hence the solution exists for all $t > t_0$. A similar argument can be given for $t_1 < t_0$.

The proof of existence and uniqueness is the same as the proof used for Theorem 4.1, except that a different metric will be used. Let M be the set of all continuous functions on $[t_0, t_1]$. Define, for $x, y \in M$,

$$\rho(x, y) = \max_{t_0 \leq t \leq t_1} e^{-L(t - t_0)} |x(t) - y(t)|,$$

where $L > K$. The proof that this is a metric space is left as an exercise. Define

$$(Ty)(t) = y_0 + \int_{t_0}^{t} f(s, y(s)) \, ds. \tag{4.3}$$

Since $f(t, y)$ is continuous, T maps functions from M to M. A fixed point of T satisfies

$$y(t) = y_0 + \int_{t_0}^{t} f(s, y(s)) \, ds,$$

which is equivalent to a solution of (4.1), (4.2). We need, then, only to show that T is a contraction mapping.

If $x(t), y(t) \in M$,

$$|(Tx)(t) - (Ty)(t)| = \left| \int_{t_0}^{t} [f(s, x(s)) - f(s, y(s))] \, ds \right|$$

$$\leq \left| \int_{t_0}^{t} |f(s, x(s)) - f(s, y(s))| \, ds \right|$$

$$\leq K \left| \int_{t_0}^{t} |x(s) - y(s)| \, ds \right| = K \int_{t_0}^{t} |x(s) - y(s)| \, ds.$$

Multiplying both sides by $e^{-L(t-t_0)}$ and rewriting slightly,

$$e^{-L(t-t_0)}|Tx(t) - Ty(t)|$$

$$\leq Ke^{-L(t-t_0)}\int_{t_0}^{t}e^{-L(s-t_0)}e^{+L(s-t_0)}|x(s) - y(s)|\,ds.$$

Since

$$e^{-L(s-t_0)}|x(s) - y(s)| \leq \max_{t_0 \leq s \leq t_1}e^{-L(s-t_0)}|x(s) - y(s)| = \rho(x,y),$$

this may be rewritten as

$$e^{-L(t-t_0)}|Tx(t) - Ty(t)| \leq K\rho(x,y)e^{-L(t-t_0)}\int_{t_0}^{t}e^{L(s-t_0)}\,ds.$$

Performing the integration yields

$$e^{-L(t-t_0)}|Tx(t) - Ty(t)| \leq \frac{K}{L}\rho(x,y)e^{-L(t-t_0)}[e^{L(t-t_0)} - 1]$$

$$\leq \frac{K}{L}[1 - e^{-L(t_1-t_0)}]\rho(x,y)$$

$$< \frac{K}{L}\rho(x,y) = \alpha\rho(x,y)$$

where $\alpha = K/L < 1$. If

$$e^{-L(t-t_0)}|Tx(t) - Ty(t)| \leq \alpha\rho(x,y)$$

for every $t \in [t_0, t_1]$, then so is the maximum over this interval. Hence,

$$\rho(Tx, Ty) \leq \alpha\rho(x,y)$$

and T is a contraction. An application of Theorem 3.1 yields a fixed point of T and hence a solution of the initial value problem (4.1), (4.2) on $[t_0, t_1]$. This contradicts the supposition that t_1 was a point that could not be reached by the extension procedure. Hence a solution exists on $[t_0, t_1]$ for every t_1, or a solution exists on $[t_0, \infty)$.

It is important to note that $f(t, y) = a(t)y + b(t)$ is uniformly Lipschitzian for t in any interval $[t_0, t_1]$ if $a(t)$ and $b(t)$ are continuous. The argument above will yield a global existence theorem for a linear equation with the observation that if there is an interval $[t_0, t_1]$ on which no solution can be continued, $|a(t)|$ and $|b(t)|$ are bounded on $[t_0, t_1]$ (since they are continuous there). Then the contraction mapping works as before, yielding a contradiction.

COROLLARY 4.3

If $a(t)$, $b(t)$ are continuous on R, there exists a unique solution $y(t)$ of

$$y' = a(t)y + b(t)$$

$$y(t_0) = y_0$$

defined for all real t.

The Lipschitz condition that was so useful in establishing the existence theorem can also be used to show how the solution changes with the initial condition. To facilitate this, the following simple lemma (a form of Gronwall's inequality) is useful.

LEMMA 4.4

Let $\varphi(t)$ be a nonnegative function that satisfies

$$\varphi(t) \le C + K \int_{t_0}^{t} \varphi(s) \, ds, \qquad t > t_0, \tag{4.4}$$

where C and K are constants, $K \ge 0$, $C > 0$. Then

$$\varphi(t) \le C e^{K(t-t_0)}, \qquad t > t_0.$$

Proof: If $\varphi(t)$ satisfies (4.4) and $K \ge 0$, $C > 0$, then

$$\frac{K\varphi(t)}{C + K \int_{t_0}^{t} \varphi(s) \, ds} \le K.$$

Integrating both sides from t_0 to $t > t_0$ yields

$$\ln\left(C + K \int_{t_0}^{s} \varphi(s) \, ds \right)\Bigg|_{t_0}^{t} \le K(t - t_0)$$

or

$$C + K \int_{t_0}^{t} \varphi(s) \, ds \le C e^{K(t-t_0)}.$$

Equation (4.4) then yields

$$\varphi(t) \le C e^{K(t-t_0)}.$$

A similar lemma holds if $t < t_0$.

Suppose that $\varphi(t)$ satisfies

$$\varphi'(t) = f(t, \varphi(t))$$
$$\varphi(t_0) = \alpha$$

(4.5)

and that $\psi(t)$ satisfies

$$\psi'(t) = f(t, \psi(t))$$
$$\psi(t_0) = \beta.$$

(4.6)

Then, using the equivalent integral equation formulation, we have

$$\varphi(t) = \alpha + \int_{t_0}^{t} f(s, \varphi(s)) \, ds$$

$$\psi(t) = \beta + \int_{t_0}^{t} f(s, \psi(s)) \, ds,$$

and so

$$\varphi(t) - \psi(t) = \alpha - \beta + \int_{t_0}^{t} (f(s, \varphi(s)) - f(s, \psi(s))) \, ds.$$

Taking absolute values yields that

$$|\varphi(t) - \psi(t)| \le |\alpha - \beta| + \left| \int_{t_0}^{t} |f(s, \varphi(s)) - f(s, \psi(s))| \, ds \right|$$

$$\le |\alpha - \beta| + K \left| \int_{t_0}^{t} |\varphi(s) - \psi(s)| \, ds \right|.$$

Application of Lemma 4.4 to the function $|\varphi(t) - \psi(t)|$ gives

$$|\varphi(t) - \psi(t)| \le |\alpha - \beta| e^{K|t - t_0|}.$$

(4.7)

Now we replace (4.6) by

$$\psi'(t) = f(t, \psi)$$
$$\psi(t_0) = \alpha_n,$$

(4.8)

where $\alpha_n \to \alpha$. We denote the solution of (4.8) by $\psi_n(t)$ and restrict our attention to a fixed interval $[t_0, t_0 + T]$. Then (4.7) tells us that

$$|\varphi(t) - \psi_n(t)| \leq |\alpha - \alpha_n| e^{K|t - t_0|}$$

$$\leq |\alpha - \alpha_n| e^{KT}.$$

Thus as $n \to \infty$, $|\varphi(t) - \psi_n(t)| \to 0$, since e^{KT} is a fixed constant that is multiplied by a quantity that tends to zero. The conclusion is that *small changes* ($|\alpha - \alpha_n|$) *in the initial condition produce small changes in the solution.*

This fact can be given a particularly elegant form using the notion of metric spaces. Let M_1 and M_2 be metric spaces with metrics ρ_1, ρ_2, respectively, and let $T : M_1 \to M_2$. T is said to be *continuous at* $x_0 \in M_1$ if for $\varepsilon > 0$ there is a $\delta > 0$ such that $\rho_2(Tx, Tx_0) < \varepsilon$ if $\rho_1(x, x_0) < \delta$. (This definition was given in Section 2 with $M_1 = M_2$.) Let M_1 be the set of real numbers with metric $|x - y|$ and let M_2 be the space $C[a, b]$ (with the usual metric). Define $T : R \to C[a, b]$ by $Ty_0 = \varphi$, where $\varphi(t)$ is the solution of

$$y' = f(t, y)$$

$$y(t_0) = y_0.$$

Since y_0 changes, Ty_0 is the solution of a different initial value problem.

THEOREM 4.5

    T is a continuous mapping.

Proof: Let $\varepsilon > 0$. Choose $\delta < \varepsilon / e^{K(b-a)}$. Then if $|y_0 - y_1| < \delta$, (4.7) yields

$$|(Ty_0)(t) - (Ty_1)(t)| \leq |y_0 - y_1| e^{K(b-a)}$$

$$< \varepsilon, \qquad \text{for every } t \in [a, b].$$
$$(4.9)$$

Since $Ty_0 - Ty_1$ is a continuous function of t, the maximum on the left-hand side occurs on the interval $[a, b]$, so

$$\rho(Ty_0, Ty_1) < \varepsilon,$$

which completes the proof.

This result gives the continuous dependence of the solution on initial conditions. In applied problems, the initial condition is often a measured quantity. This theorem gives the reassurance that a small error in this measurement produces a small error in the solution. Without this fact, differential equation models would have no predictive value.

It is possible to prove a similar result as the initial time, t_0, varies. With great care it is also possible to let the function $f(t, y)$ vary. We will not deal with these properties.

EXERCISES

1. Determine if the following functions satisfy a local or a uniform Lipschitz condition.

   (a) $|y|$ (b) $t^2|y|$

   (c) $\sin(y)$ (d) $\tan^{-1}(y)$

   (e) te^y (f) e^{-y}

   (g) $\dfrac{t^2 y}{1 + y^2}$

2. Determine the maximum interval of existence for the following differential equations by solving. Then fix a rectangle R about the given initial condition, determine M, and compare with the estimate given by Theorem 4.1 (i.e., compute α of Theorem 4.1).

   (a) $y' = 1 + y^{-2}$ (b) $y' = -y$ (c) $y' = \dfrac{1}{1 + y^2}$

        $y(0) = 1$ $y(0) = 1$ $y(0) = 0$

3. The function $f(y) = 1 + y^2$ is locally Lipschitzian. Consider $y' = 1 + y^2$, $y(0) = 0$, and construct an appropriate rectangle Ω as in the hypothesis for Theorem 4.1. Compute α. Take the new initial condition $y(t_0 + \alpha)$, find a new rectangle, and repeat the process.

4. Let $\varphi(t)$ be a nonnegative function that satisfies

$$\varphi(t) \leq C + \int_{t_0}^{t} K(s)\varphi(s)\,ds$$

   where $C > 0$ and $K(s) \geq 0$. Show that

$$\varphi(t) \leq Ce^{\int_{t_0}^{t} K(s)\,ds}, \qquad t \geq t_0.$$

5. What conclusion may be made if $C = 0$ in Exercise 4? (Note: the proof you gave in Exercise 4 may fail if $C = 0$.)

5. The Initial Value Problem for Systems of Differential Equations

We now turn our attention to the system of nonlinear differential equations

$$y_1' = f_1(t, y_1, \ldots, y_n)$$
$$y_2' = f_2(t, y_1, \ldots, y_n)$$
$$\vdots \qquad\qquad\qquad\qquad (5.1)$$
$$y_n' = f_n(t, y_1, \ldots, y_n)$$

with initial conditions

$$y_1(t_0) = \alpha_1$$
$$y_2(t_0) = \alpha_2$$
$$\vdots$$
$$y_n(t_0) = \alpha_n. \tag{5.2}$$

If y is the vector with components $y_1, \ldots, y_n$, $f(t, y)$ is a vector function with components $f_1(t, y_1, y_2, \ldots, y_n), f_2(t, y_1, y_2, \ldots, y_n), \ldots, f_n(t, y_1, y_2, \ldots, y_n)$, and y_0 is a vector with components $\alpha_1, \ldots, \alpha_n$, this problem can be written in vector notation as

$$y' = f(t, y) \tag{5.3}$$

$$y(t_0) = y_0. \tag{5.4}$$

A *solution* of (5.3), (5.4) on an interval I is a differentiable, vector-valued function $\varphi(t)$ such that $(t, \varphi(t))$ is in the domain of definition of f and such that

$$\varphi'(t) = f(t, \varphi(t))$$

for $t \in I$ and

$$\varphi(t_0) = y_0.$$

If x is a vector in R^n with components $x_1, \ldots, x_n$, we continue the convention that

$$|x| = |x_1| + |x_2| + \cdots + |x_n| = \sum_{i=1}^{n} |x_i|.$$

A vector-valued function $f(t, y)$ is said to satisfy a *Lipschitz condition in a region G* if for every (t, x), (t, y) in G,

$$|f(t, x) - f(t, y)| \le K|x - y| \tag{5.5}$$

for some constant K. To relate this to the Lipschitz constant for scalar functions, note that if each component $f_i(t, y)$ satisfies a Lipschitz condition with respect to each component of y_i with Lipschitz constant L_i, then

$$|f(t, x) - f(t, y)| = |f_1(t, x) - f_1(t, y)| + |f_2(t, x) - f_2(t, y)|$$
$$+ \cdots + |f_n(t, x) - f_n(t, y)|, \tag{5.6}$$

and for each $i = 1, \ldots, n$,

$$
\begin{aligned}
|f_i(t, x) - f_i(t, y)| &= |f_i(t, x_1, \ldots, x_n) - f_i(t, y_1, \ldots, y_n)| \\
&= |f_i(t, x_1, x_2, \ldots, x_n) - f_i(t, y_1, x_2, \ldots, x_n) \\
&\quad + f_i(t, y_1, x_2, \ldots, x_n) - f_i(t, y_1, y_2, x_3, \ldots, x_n) \\
&\quad + \cdots + f_i(t, y_1, \ldots, y_{n-1}, x_n) - f(t, y_1, \ldots, y_n)| \\
&\le L_1^i |x_1 - y_1| + L_2^i |x_2 - y_2| + \cdots + L_n^i |x_n - y_n|.
\end{aligned}
$$

Let $L^i = \max [L_1^i, L_2^i, \ldots, L_n^i]$, then this becomes

$$
\begin{aligned}
|f_i(t, x) - f_i(t, y)| &\le L^i[|x_1 - y_1| + \cdots + |x_n - y_n|] \\
&= L^i |x - y|.
\end{aligned}
$$

If $K = \sum_{i=1}^n L^i$, then (5.6) yields $|f(t, x) - f(t, y)| \le K|x - y|$. Thus if each component of $f(t, y)$ satisfies a Lipschitz condition with respect to each component of y, then $f(t, y)$ satisfies the vector-type Lipschitz condition (5.5).

Let $C_n[a, b]$ be the set of all n-vector-valued functions defined on $[a, b]$ with metric ρ defined by

$$
\begin{aligned}
\rho(x, y) &= \max_{a \le t \le b} |x(t) - y(t)| \\
&= \max_{a \le t \le b} \sum_{i=1}^n |x_i(t) - y_i(t)|,
\end{aligned}
$$

where $x_i(t)$, $y_i(t)$ are the components of x and y, $x, y \in C_n[a, b]$. We leave as an exercise the fact that this definition of $\rho(x, y)$ makes $C_n[a, b]$ a complete metric space. We use this metric space to prove the following existence theorem.

THEOREM 5.1

   Let $f(t, y)$ be continuous on $\Omega \{(t, y) | |t - t_0| \le a, |y - y_0| \le b\}$ and satisfy a Lipschitz condition there, with Lipschitz constant K. Let $M = \max |f(t, y)|$ for $(t, y) \in \Omega$. There exists a unique solution of

$$
y' = f(t, y)
$$

$$
y(t_0) = y_0
$$

valid on $|t - t_0| \le \alpha < \min [a, b/M, 1/K]$.

The proof of this theorem follows that of Theorem 4.1 but requires a few preliminary comments. First of all, it is easy to see that the existence of a unique solution is equivalent to the existence of a unique continuous solution of the vector integral equation

$$y(t) = y_0 + \int_{t_0}^{t} f(s, y(s)) \, ds. \tag{5.7}$$

The arguments are formally the same as those in Theorem 2.1. We also need to find an appropriate space to apply the contraction mapping theorem. When this is done, the proof of Theorem 5.1 is formally the same as for Theorem 4.1—the symbols have changed their meaning but the structure of the proof is the same. This is part of the power of abstract mathematical tools.

Let B_n denote the subset of $C_n[t_0 - \alpha, t_0 + \alpha]$ whose elements satisfy $\rho(y, y_0) \le b$. That B_n is a complete metric space will follow, as before, from the fact that the metric space $C_n[t_0 - \alpha, t_0 + \alpha]$ is complete. The completeness of $C_n[t_0 - \alpha, t_0 + \alpha]$ will follow from the fact that each component of an element of this space is itself an element of the complete metric space $C[t_0 - \alpha, t_0 + \alpha]$ and that the number of components is finite. The proof of completeness is left as an exercise.

Proof of Theorem 5.1:   Define a mapping on B_n by

$$(Ty)(t) = y_0 + \int_{t_0}^{t} f(s, y(s)) \, ds.$$

We need first to show that T maps B_n into B_n. Since $f(t, \varphi(t))$ is continuous, T certainly maps B_n into $C_n[t_0 - \alpha, t_0 + \alpha]$. If $\varphi \in B$, then

$$|(T\varphi)(t) - y_0| = \left| \int_{t_0}^{t} f(s, \varphi(s)) \, ds \right|$$

$$\le \left| \int_{t_0}^{t} |f(s, \varphi(s))| \, ds \right|$$

$$\le M|t - t_0| \le M\alpha \le b$$

by the way α was defined. By what is now a standard argument, $|(T\varphi)(t) - y_0|$ is a continuous function on $[t_0 - \alpha, t_0 + \alpha]$ and assumes its maximum. Thus

$$\rho(T\varphi, y_0) \le b$$

for $\varphi \in B$ and T maps B_n into B_n.

The existence of a fixed point of T will follow from the contraction mapping theorem. If $x, y \in B_n$, then

$$|Tx(t) - Ty(t)| = \left| \int_{t_0}^{t} [f(s, x(s)) - f(s, y(s))] \, ds \right|$$

$$\leq \left| \int_{t_0}^{t} |f(s, x(s)) - f(s, y(s))| \, ds \right|$$

$$\leq \left| \int_{t_0}^{t} K |x(s) - y(s)| \, ds \right|.$$

Since $|x(s) - y(s)| \leq \rho(x, y)$,

$$|Tx(t) - Ty(t)| \leq K\rho(x, y) |t - t_0|$$
$$\leq K\alpha\rho(x, y).$$

Since this holds for every t, we can take the maximum over $[t_0 - \alpha, t_0 + \alpha]$ and obtain

$$\rho(Tx, Ty) \leq K\alpha\rho(x, y).$$

Since $K\alpha < 1$, T is a contraction and the existence of a unique fixed point of T follows from the contraction mapping theorem. This unique fixed point is the only continuous solution of the integral equation (5.7) and hence is a unique solution of (5.3), (5.4).

If $f(t, y)$ satisfies a Lipschitz condition with the same constant for all pairs (t, x), (t, y), then a global existence theorem can be proved for systems.

THEOREM 5.2

Let $f(t, y)$ be continuous in t and y and satisfy a uniform Lipschitz condition for every pair (t, x), (t, y). Then there exists a unique solution of (5.3), (5.4) valid for all t.

The proof follows that of Theorem 4.2 and is left as an exercise.

Continuity with respect to initial conditions also holds for systems. Consider the set of initial value problems

$$y' = f(t, y)$$
$$y(t_0) = \alpha_k.$$
(5.8)

THEOREM 5.3

Let $f(t, y)$ be continuous in t and Lipschitzian in y. Fix $I = [t_0, T]$ for any finite T and let φ_{α_k} denote the solution of (5.8) viewed as an element of $C_n[t_0, T]$, $k = 1, 2, \ldots$. If $\alpha_k \to \alpha$, then $\rho(\varphi_{\alpha_k}, \varphi_\alpha) \to 0$.

In particular, for any fixed $t \in [t_0, T]$, $\varphi_{\alpha_k}(t) \to \varphi_\alpha(t)$ as $k \to \infty$. The proof is left as an exercise.

Sometimes the right-hand side of the equation depends on a parameter; for example,

$$x' = f(t, x, \lambda)$$
$$x(t_0) = x_0,$$

(5.9)

where λ is a real number. For each different value of the parameter λ, (5.9) represents a different system of differential equations. Yet if f is continuous in λ as well as in t and x, and if solutions of (5.9) are unique for each choice of λ, we anticipate that the solution of (5.9) will vary smoothly.

The following theorem makes this more precise.

THEOREM 5.4

Let $f(t, y, \lambda)$ be continuous in t and λ and satisfy a uniform Lipschitz condition in y. Let φ_λ denote the solution of

$$y' = f(t, y, \lambda)$$
$$y(t_0) = x_0$$

(5.10)

viewed as an element of $C_n[t_0, T]$, some finite T. If $\lambda_k \to \lambda_0$, then

$$\rho(\varphi_{\lambda_k}, \varphi_{\lambda_0}) \to 0$$

as $k \to \infty$.

Proof: The proof proceeds by a small trick that raises the order of the system by one and reduces the problem to the previous case, Theorem 5.3. Instead of (5.10), consider the $(n + 1)$-dimensional problem

$$y' = f(t, y, \lambda)$$
$$\lambda' = 0$$
$$y(t_0) = x_0$$
$$\lambda(t_0) = \lambda_0.$$

(5.11)

By Theorem 5.3, solutions of (5.11) vary continuously with respect to their initial conditions. The equation $\lambda' = 0$ makes λ a constant, and $\lambda(t_0) = \lambda_0$ fixes that value. Thus continuity with respect to the initial condition λ_0 in (5.11) is continuity with respect to the parameter λ in (5.10).

We now use the existence theorem for systems to provide the result promised earlier for scalar nth-order differential equations. Although only linear equations were considered in Chapter 1, we state the theorem for the general case. The differential equation takes the form

$$y^{(n)} + f(t, y, y', y'', \ldots, y^{(n-1)}) = 0, \tag{5.12}$$

where by $y^{(i)}$ we mean the ith derivative of the function y. The appropriate initial conditions (as we shall prove) are

$$
\begin{aligned}
y(t_0) &= \alpha_0 \\
y'(t_0) &= \alpha_1 \\
y''(t_0) &= \alpha_2 \\
&\vdots \\
y^{(n-1)}(t_0) &= \alpha_{n-1},
\end{aligned}
\tag{5.13}
$$

where $\alpha_0, \alpha_1, \ldots, \alpha_{n-1}$ are real numbers. The appropriate Lipschitz condition for one function of several variables takes the form

$$
\begin{aligned}
&|f(t, y, y', \ldots, y^{(n-1)}) - f(t, z, z', \ldots, z^{(n-1)})| \\
&\quad \le K_0|y - z| + K_1|y' - z'| + \cdots + K_{n-1}|y^{(n-1)} - z^{(n-1)}|.
\end{aligned}
\tag{5.14}
$$

THEOREM 5.5

Let $f(t, y, y', \ldots, y^{(n-1)})$ be continuous on the set $\Omega = \{(t, y, y', \ldots, y^{(n-1)})\,||\,t - t_0| \le a, \quad |y - \alpha_0| \le b_0, \quad |y' - \alpha_1| \le b_1, \quad \ldots, |y^{(n-1)} - \alpha_{n-1}| \le b_{n-1}\}$, and satisfy the Lipschitz condition (5.14) there. Then there exists a unique solution of (5.12) satisfying the initial condition (5.13), valid for $t_0 \le t \le t_1$, for some number t_1, $t_0 < t_1 \le t_0 + a$.

We can be more precise about the length of the existence interval, but it is not necessary to do so. The important point is that the interval is nonempty.

Proof of Theorem 5.5. Change (5.12) into a system of the form

$$y_1' = y_2$$
$$y_2' = y_3$$
$$\vdots \tag{5.15}$$
$$y_n' = -f(t, y_1, y_2, \ldots, y_n)$$

with initial conditions

$$y_1(t_0) = \alpha_0$$
$$y_2(t_0) = \alpha_1$$
$$\vdots \tag{5.16}$$
$$y_n(t_0) = \alpha_{n-1}.$$

The Lipschitz condition (5.14) can be put into the form (5.7) if we let $K = 1 + \max[K_0, K_1, \ldots, K_{n-1}]$. Theorem 5.1 now applies in a neighborhood of the vector $(\alpha_0, \alpha_1, \ldots, \alpha_{n-1})$. Since the problem (5.15), (5.16) is equivalent to (5.12), (5.13), the proof is complete.

EXERCISES

1. Show that $C_n[a, b]$ is a metric space.

2. Show that $C_n[a, b]$ is complete.

3. Prove that a solution of the system (5.3), (5.4) exists if and only if there is a continuous solution of the vector integral equation (5.7).

4. Prove Theorem 5.2.

5. Prove Theorem 5.3.

6. Replace λ by a vector $(\lambda_1, \ldots, \lambda_p)$ and reprove Theorem 5.4.

7. Show formally that (5.15), (5.16) is equivalent to (5.12), (5.13). (*Hint:* Mimic the linear case from Chapter 1.)

6. An Existence Theorem for a Boundary Value Problem

The contraction mapping theorem developed in Section 3 makes it possible to give a simple proof of an existence and uniqueness theorem for some second-

order nonlinear boundary value problems. Specifically, the existence and uniqueness of solutions of the boundary value problem

$$y'' + f(t, y) = 0 \tag{6.1}$$

$$y(a) = 0 \tag{6.2}$$

$$y(b) = 0 \tag{6.3}$$

will be established in this section. The boundary conditions (6.2), (6.3) appear
quite restrictive, but the conditions on f are such that other boundary conditions
can be put into this form (see the exercises at the end of the section).

To use the contraction mapping theorem, an appropriate metric space and
an appropriate operator must be chosen. The correct space is $C[a, b]$, the space of
functions that are continuous on $[a, b]$ with the metric $\rho(x, y) = \max_{[a,b]} |x(t) -
y(t)|$. If the function in (6.1) were $f(t, y, y')$, a different choice would have to be
made—see the exercises at the end of the section. The choice of an operator T
such that a fixed point of T is a solution of (6.1)–(6.3) is more difficult. To
motivate the proper choice of operator would take us into a lengthy side
discussion, so, since we are interested only in exhibiting the application of the
contraction mapping theorem, we state it directly and then verify that our choice
is the correct one.

We first define a function of two variables on $[a, b] \times [a, b]$ by

$$G(t, \tau) = \begin{cases} \dfrac{(t - a)(b - \tau)}{b - a}, & a \leq t \leq \tau \leq b \\ \dfrac{(\tau - a)(b - t)}{b - a}, & a \leq \tau \leq t \leq b. \end{cases} \tag{6.4}$$

Hereafter we refer to $G(t, \tau)$ as the *Green's function* for (6.1)–(6.3). A more
general treatment will be given in the next chapter. Note that $G(t, \tau)$ is a
continuous function.

The integral equation that plays a role similar to that of (2.3) for initial
value problems is

$$y(t) = \int_a^b G(t, \tau) f(\tau, y(\tau)) \, d\tau. \tag{6.5}$$

LEMMA 6.1

   **A continuous solution of (6.5) is a solution of (6.1)–(6.3). Every solution of
   (6.1)–(6.3) is a solution of (6.5).**

Proof: Let $\varphi(t)$ be a continuous function that solves (6.5). Since $G(a, \tau) = G(b, \tau) = 0$ for every $\tau \in [a, b]$, $\varphi(a) = \varphi(b) = 0$ and (6.2) and (6.3) are satisfied. To show that $\varphi(t)$ satisfies the differential equation is an exercise in differentiating under the integral sign. To facilitate this, rewrite (6.5) as

$$\varphi(t) = \int_a^t G(t, \tau) f(\tau, \varphi(\tau)) \, d\tau + \int_t^b G(t, \tau) f(\tau, \varphi(\tau)) \, d\tau.$$

Then, since $G(t, \tau)$ is continuous,

$$\varphi'(t) = G(t, t) f(t, \varphi(t)) + \int_a^t G_t(t, \tau) f(\tau, \varphi(\tau)) \, d\tau$$

$$- G(t, t) f(t, \varphi(t)) + \int_t^b G_t(t, \tau) f(\tau, \varphi(\tau)) \, d\tau$$

$$= \int_a^b G_t(t, \tau) f(\tau, \varphi(\tau)) \, d\tau,$$

where $G_t(t, \tau) = \partial/\partial t[G(t, \tau)]$. Since $G_t(t, \tau)$ is not continuous (there is a jump across $t = \tau$), care must be exercised in taking the derivative. In particular, we must take limits from the appropriate side, since

$$\lim_{\substack{\tau \to t \\ \tau < t}} G_t(t, \tau) = -\frac{t - a}{b - a}$$

while

$$\lim_{\substack{\tau \to t \\ \tau > t}} G_t(t, \tau) = \frac{b - t}{b - a}.$$

Note that since $G(t, \tau)$ is linear in t (see (6.4)) for each τ, $G_{tt}(t, \tau) = 0$. Hence

$$\varphi''(t) = \int_a^t G_{tt}(t, \tau) f(\tau, \varphi(\tau)) \, d\tau + \lim_{\substack{\tau \to t \\ \tau < t}} G_t(t, \tau) f(\tau, \varphi(\tau))$$

$$+ \int_t^b G_{tt}(t, \tau) f(\tau, \varphi(\tau)) \, d\tau - \lim_{\substack{\tau \to t \\ \tau > t}} G_t(t, \tau) f(\tau, \varphi(\tau))$$

$$= \left(-\frac{t - a}{b - a} - \frac{b - t}{b - a} \right) f(t, \varphi(t))$$

$$= -f(t, \varphi(t)).$$

Thus $\varphi''(t) + f(t, \varphi(t)) = 0$, which proves the first statement of the lemma. Proof of the second is left as an exercise.

An existence and uniqueness theorem for (6.1)–(6.3) can now be proved by showing that (6.5) has a unique, continuous solution. To apply the contraction mapping theorem, it is necessary to assume that f satisfies a uniform Lipschitz condition on $[a, b] \times R$, specifically that there is a number K such that, for every pair of points (t, y_1), (t, y_2),

$$|f(t, y_1) - f(t, y_2)| \leq K|y_1 - y_2|. \tag{6.6}$$

THEOREM 6.2

Let $f(t, y)$ be continuous on $[a, b] \times R$ and satisfy a Lipschitz condition there. If $b - a$ is so small that $K(b - a)^2/4 < 1$, where K is the Lipschitz constant in (6.6), then there exists a unique solution of

$$y'' + f(t, y) = 0$$

$$y(a) = 0$$

$$y(b) = 0.$$

Proof: Define a mapping T on $C[a, b]$ into $C[a, b]$ by

$$(T\varphi)(t) = \int_a^b G(t, \tau) f(\tau, \varphi(\tau)) \, d\tau, \tag{6.7}$$

where $G(t, \tau)$ is defined by (6.4). As noted above, a fixed point of this operator would be a solution to the boundary value problem. We check then to see if T is a contraction mapping:

$$|T\varphi_1(t) - T\varphi_2(t)| = \left| \int_a^b G(t, \tau) f(\tau, \varphi_1(\tau) \, d\tau \right.$$

$$\left. - \int_a^b G(t, \tau) f(\tau, \varphi_1(\tau)) \, d\tau \right|$$

$$\leq \int_a^b |G(t, \tau)| |f(\tau, \varphi_1(\tau)) - f(\tau, \varphi_2(\tau))| \, d\tau.$$

In view of the Lipschitz condition, we have

$$|T\varphi_1(t) - T\varphi_2(t)| \leq K \int_a^b |G(t, \tau)| |\varphi_1(\tau) - \varphi_2(\tau)| \, d\tau$$

$$\leq K\rho(\varphi_1, \varphi_2) \int_a^b |G(t, \tau)| \, d\tau.$$

Before proceeding further, we need an estimate on the integral involving $G(t, \tau)$.

LEMMA 6.3
  If $G(t, \tau)$ is defined by (6.4), then

$$\int_a^b |G(t, \tau)| \, d\tau \le \frac{(b - a)^2}{4}.$$

Proof: First of all, from (6.4), $G(t, \tau) \ge 0$, and since $\partial G/\partial t > 0$ if $t \le \tau$ and $\partial G/\partial t < 0$ if $t \ge \tau$, the maximum value of $G(t, \tau)$ must occur at $\tau = t$. To find this value, differentiate $G(t, t)$ with respect to t to obtain the occurrence of the maximum as a solution of

$$(b - t) + (t - a)(-1) = 0$$

or

$$t = \frac{b + a}{2}.$$

Putting this value into $G(t, t)$ yields

$$G\left(\frac{b + a}{2}, \frac{b + a}{2}\right) = \frac{\left(\dfrac{b + a}{2} - a\right)\left(b - \dfrac{b + a}{2}\right)}{b - a}$$

$$= \frac{(b - a)^2}{4(b - a)} = \frac{(b - a)}{4}.$$

Thus $|G(t, \tau)| \le (b - a)/4$. An integration immediately produces the estimate of the lemma,

$$\int_a^b |G(t, \tau)| \, d\tau \le \frac{(b - a)^2}{4}.$$

(With a little more effort we can obtain $(b - a)^2/8$; see the exercises.)

Returning now to our investigation of T, we see that

$$|T\varphi_1(t) - T\varphi_2(t)| \le K\rho(\varphi_1, \varphi_2)\frac{(b - a)^2}{4}.$$

Finally, taking the maximum for t in $[a, b]$, we have

$$\rho(T\varphi_1, T\varphi_2) \le K\frac{(b - a)^2}{4}\rho(\varphi_1, \varphi_2).$$

Thus T is a contraction if

$$K\frac{(b-a)^2}{4} < 1,$$

which is a hypothesis of the theorem. Hence T has a unique fixed point, so that (6.5) has a unique, continuous solution and, finally, (6.1)–(6.3) has a unique solution.

EXERCISES

1. Show that the existence and uniqueness of a solution of $y'' + f(t, y) = 0$, $y(a) = A$, $y(b) = B$ for f satisfying a Lipschitz condition is equivalent to the existence and uniqueness of a solution of $y'' + f(t, y)$, $y(a) = 0$, $y(b) = 0$. (*Hint:* Use the function $\ell(t) = [bA - aB + (B - A)t]/(b - a)$ to change this problem into the one in Theorem 6.2.)

2. Show that $G(t, \tau)$ is continuous and find regions where $G(t, \tau)$ is differentiable.

3. Show that a solution of (6.1)–(6.3) is a solution of (6.5).

4. Improve Theorem 6.2 by weakening the hypothesis to $K(b - a)^2/8 < 1$. (*Hint:* Integrate $G(t, \tau)$.)

5. Let M be the set of functions defined on $[a, b]$ that have a continuous derivative. Define

$$\rho_{K,L}(x, y) = \max_{[a,b]} [K|x(t) - y(t)| + L|x'(t) - y'(t)|],$$

where $K > 0$ and $L > 0$. Show that $(M, \rho_{K,L})$ forms a metric space.

6. Show that the existence of a solution of the boundary value problem

$$y'' + f(t, y, y') = 0$$

$$y(a) = y(b) = 0$$

is equivalent to the existence of a continuous solution of a pair of integral equations

$$y(t) = \int_a^b G(t, \tau) f(\tau, y(\tau), y'(\tau)) \, d\tau$$

$$y'(t) = \int_a^b G_t(t, \tau) f(\tau, y(\tau), y'(\tau)) \, d\tau.$$

7. Suppose that f satisfies a Lipschitz condition of the form $|f(t, x_1, y_1) - f(t, x_2, y_2)| \le K|x_1 - x_2| + L|y_1 - y_2|$. Use the metric space of Exercise 5 and the contraction mapping principle to show that the boundary value problem in Exercise 6 has a unique solution, provided

$$\frac{K(b-a)^2}{8} + L\frac{(b-a)}{2} < 1.$$

(*Hint:* Show that $\int_a^b |G_t(t,\tau)|\, d\tau \le (b-a)/2$.)

8. Complete the proof of Lemma 6.1. (*Hint:* Multiply both sides of the equation by $G(t,t)$ and integrate by parts.)

REFERENCES

Material on initial value problems appears in most intermediate and advanced textbooks on ordinary differential equations. A very readable account, for example, is in Chapter 1 of

Coppel, W. A. *Stability and Asymptotic Behavior of Differential Equations.* Lexington, MA: D. C. Heath & Company, 1965.

More information on two-point boundary value problems and on the application of the contraction mapping theorem can be found in

Bailey, P. B., L. F. Shampine, and P. Waltman. *Nonlinear Two Point Boundary Value Problems.* New York: Academic Press, 1968.

4

Boundary Value Problems

1. Introduction

An elementary course in ordinary differential equations often leaves the impression that only initial value problems are of interest. Indeed, all of the material thus far, except for one existence theorem in Chapter 3, has concentrated on initial value problems. There are, however, a variety of other possible conditions that are important in applications. For initial value problems, all of the conditions were placed at one value of the independent variable—the initial "time." When conditions occur for different values of the independent variable, the resulting problem is called a *boundary value problem*. Often the independent variable is space rather than time, and the conditions correspond to limits placed on the motion or other aspects of the object being modeled—the ends of a rod may be fixed, temperature at a point may be held constant, and so on. In this chapter the focus is on boundary value problems and on one class—the Sturm-Liouville problems—in particular.

The chapter begins with a consideration of second-order linear boundary value problems and develops the general structure of such problems. The idea of a "Green's function" is introduced, and inhomogeneous problems are treated in these terms. The key question for the existence of such functions turns out to be the distance between zeros of certain solutions. Some useful theorems, oscillation and comparison theorems, are discussed, that help answer this question. The Sturm-Liouville problem, a type of boundary value problem that contains a parameter, is introduced and a geometric argument, which has much in common with the arguments found in Chapter 2, is used to establish the existence of very special values of the parameter—eigenvalues. Properties of the solutions corresponding to these choices of the parameter are developed and used to expand other functions in series that are particularly useful in applications. The use of the Sturm-Liouville theory is illustrated with some traditional physical applications. Finally, a nonlinear eigenvalue problem is discussed.

2. Linear Boundary Value Problems

We begin by considering a linear second-order differential equation with variable coefficients and with conditions placed on the solution at two different values of the independent variable. Although the theory of linear boundary value problems can be developed in this generality, computations are really possible at this level only when the coefficients are constant. There is a corresponding theory for general linear systems, such as those considered in Chapter 1, but it is too difficult to be considered in this course. The problem of interest takes the form

$$y'' + a(t)y' + b(t)y = e(t) \tag{2.1}$$

$$y(t_0) = A \tag{2.2}$$

$$y(t_1) = B, \tag{2.3}$$

where $a(t)$, $b(t)$, and $e(t)$ are continuous functions and $t_1 > t_0$. Discontinuous coefficients and the case $t_1 = \infty$ are important but are not considered here. We could apply Theorem 6.2 of Chapter 3 to assert the existence and uniqueness of solutions of (2.1)–(2.3) if $t_1 - t_0$ is sufficiently small. However, since the problem is linear we are able to say much more.

It will be convenient to use an operator notation similar to that introduced in Chapter 1 for systems of equations. On the set of functions on $[t_0, t_1]$ that have two continuous derivatives, define a mapping into the set of continuous functions by

$$L[z](t) = z''(t) + a(t)z'(t) + b(t)z(t).$$

The image of L is a continuous function, since z, its derivatives, and $a(t)$ and $b(t)$ are continuous functions. Thus the set of all solutions of (2.1) is just the set of twice-differentiable functions that map, under this operator L, into the continuous function $e(t)$. We will not make any use of the operator properties but only of the notational convenience of this view. Clearly, L is a linear operator.

First consider the unforced case, the case where $e(t) = 0$ for every t. It is important to note that the following theorem changes the question of existence and uniqueness of a solution of the boundary value problem to a question of existence of a nontrivial solution with two zeros at the boundary points. This illustrates the importance of determining the locations of zeros of solutions of differential equations—a natural introduction for a topic considered later.

THEOREM 2.1

A necessary and sufficient condition that there exists a unique solution of the boundary value problem

$$y(t_0) = A \tag{2.2}$$

$$y(t_1) = B \tag{2.3}$$

$$L[y] = y'' + a(t)y' + b(t)y = 0, \tag{2.4}$$

$a(t)$, $b(t)$ continuous, is that there is no nontrivial solution of (2.4) that has a zero at t_0 and a zero at t_1.

Proof: Let $\varphi(t)$, $\psi(t)$ be two linearly independent solutions of (2.4). Such solutions exist; for example, take $\varphi(t_0) = 1$, $\varphi'(t_0) = 0$, $\psi(t_0) = 0$, $\psi'(t_0) = 1$; apply Theorem 5.5 of Chapter 3. We attempt to find constants c_1 and c_2 such that

$$c_1\varphi(t_0) + c_2\psi(t_0) = A$$
$$c_1\varphi(t_1) + c_2\psi(t_1) = B \tag{2.5}$$

holds, for then $\Omega(t) = c_1\varphi(t) + c_2\psi(t)$ is a solution of the differential equation (2.4) and satisfies the boundary conditions (2.2) and (2.3) by virtue of (2.5). There exists a unique solution c_1, c_2 of the algebraic system (2.5) if and only if the coefficient determinant is not zero, that is, if and only if

$$\varphi(t_0)\psi(t_1) - \varphi(t_1)\psi(t_0) \neq 0. \tag{2.6}$$

The proof of the theorem is thus reduced to showing that (2.6) is equivalent to there being no nontrivial solution of (2.4) with zeros at t_0 and t_1.

Since every solution is a linear combination of $\varphi(t)$ and $\psi(t)$, there is a nontrivial solution $\chi(t) = \alpha\varphi(t) + \beta\psi(t)$ with zeros at t_0 and t_1 if and only if α and β are nontrivial solutions of the algebraic system

$$\alpha\varphi(t_0) + \beta\psi(t_0) = 0$$
$$\alpha\varphi(t_1) + \beta\psi(t_1) = 0. \tag{2.7}$$

In turn, there is a unique solution of (2.7) if and only if (2.6) holds. If (2.6) holds, $\alpha = \beta = 0$ is the only solution of (2.7), and if (2.6) does not hold, α and β can be chosen so that $\chi(t)$ is nontrivial and has zeros at t_0 and t_1. This completes the proof.

Although the same technique will work for the forced equation

$$y'' + a(t)y' + b(t)y = e(t) \tag{2.8}$$

and boundary conditions

$$y(t_0) = A \tag{2.9}$$

$$y(t_1) = B, \tag{2.10}$$

we can also express the solution as a certain integral, in a way similar to that used for the boundary value problem in Chapter 3. Let $\varphi(t)$ and $\psi(t)$ be solutions of the unforced equation with $\varphi(t_0) = 0$ and $\psi(t_1) = 0$ and suppose that the Wronskian of these two solutions, $W(\varphi, \psi)$, is not zero. (Recall that the Wronskian of $\varphi(t)$ and $\psi(t)$ is given by $W(\varphi, \psi)(t) = \varphi(t)\psi'(t) - \psi(t)\varphi'(t)$.) The function

$$G(t, \tau) = \begin{cases} \dfrac{\varphi(t)\psi(\tau)}{W(\varphi, \psi)(\tau)}, & t_0 \le t \le \tau \\[2mm] \dfrac{\varphi(\tau)\psi(t)}{W(\varphi, \psi)(\tau)}, & \tau \le t \le t_1 \end{cases} \tag{2.11}$$

is called the *Green's function* for the boundary value problem (2.8)–(2.10). The theory of Green's functions and their construction are beyond the scope of this course, but we will establish certain properties needed in the proof of the main theorem. (Exercise 6 shows that $G(t, \tau)$ is independent of the choice of $\varphi(t)$ within the family of solutions through $(t_0, 0)$. A similar result is true for $\psi(t)$ and $(t_1, 0)$.)

First, $G(t, \tau)$ is continuous. There is no question except for the point $t = \tau$, and here both parts of the definition of $G(t, \tau)$ are the same. Second, there is a discontinuity in $\partial G(t, \tau)/\partial t$ at $\tau = t$. By $\partial G(t, t^+)/\partial t$ we mean $\lim_{\tau \to t \atop \tau > t} \partial G(t, \tau)/\partial t$, and if the $+$ is replaced by $-$, the limit is taken for $t > \tau$. If these two limits are different, $\partial G(t, \tau)/\partial t$ is discontinuous. Differentiating $G(t, \tau)$ with respect to t in the regions $t < \tau$ and $t > \tau$ and taking limits, we have

$$\frac{\partial G(t, t^+)}{\partial t} = \frac{\varphi'(t)\psi(t)}{W(\varphi, \psi)(t)}$$

$$\frac{\partial G(t, t^-)}{\partial t} = \frac{\varphi(t)\psi'(t)}{W(\varphi, \psi)(t)}$$

or

$$\frac{\partial G(t, t^-)}{\partial t} - \frac{\partial G(t, t^+)}{\partial t} = \frac{\varphi(t)\psi'(t) - \varphi'(t)\psi(t)}{W(\varphi, \psi)(t)}$$

$$= \frac{W(\varphi, \psi)(t)}{W(\varphi, \psi)(t)} = 1.$$

$\partial G(t, \tau)/\partial t$ has a jump of one "unit" at $\tau = t$.

Finally, we show that for a fixed τ, $G(t, \tau)$ satisfies (2.4) on the intervals $[t_0, \tau)$ and $(\tau, t_1]$. (It is necessary to split the interval since, as noted above, $\partial G(t, \tau)/\partial t$ is discontinuous at $t = \tau$. For $t_0 \le t < \tau$,

$$L[G(t, \tau)] = L\left[\frac{\varphi(t)\psi(\tau)}{W(\varphi, \psi)(\tau)}\right]$$

$$= \frac{\psi(\tau)}{W(\varphi, \psi)(\tau)} L[\varphi] = 0,$$

since $\varphi(t)$ is given to satisfy $L[\varphi] = 0$. The same kind of argument holds for $(\tau, t_1]$ with $\psi(t)$ replacing $\varphi(t)$.

THEOREM 2.2

Let $a(t)$, $b(t)$, $e(t)$ be continuous functions defined on $[t_0, t_1]$. If no nontrivial solution of (2.4) has a zero at t_0 and a zero at t_1, there exists a unique solution of the boundary value problem

$$y'' + a(t)y' + b(t)y = e(t) \tag{2.12}$$

$$y(t_0) = 0 \tag{2.13}$$

$$y(t_1) = 0. \tag{2.14}$$

Proof: Define

$$x(t) = \int_{t_0}^{t_1} G(t, \tau)e(\tau)\, d\tau.$$

We show by direct computation that $x(t)$ satisfies the differential equation (2.12) and the boundary conditions (2.13) and (2.14). First,

$$x(t_0) = \int_{t_0}^{t_1} G(t_0, \tau)e(\tau)\, d\tau.$$

Now $G(t_0, \tau) = 0$, so $x(t_0) = 0$. Similarly, $x(t_1) = 0$, since $G(t_1, \tau) = 0$. In order to facilitate differentiating $x(t)$, we break the integral at t, writing

$$x(t) = \int_{t_0}^{t} G(t, \tau)e(\tau)\, d\tau + \int_{t}^{t_1} G(t, \tau)e(\tau)\, d\tau.$$

Then

$$x'(t) = \int_{t_0}^{t} \frac{\partial}{\partial t} G(t, \tau)e(\tau)\, d\tau + G(t, t^-)e(t)$$

$$+ \int_{t}^{t_1} \frac{\partial}{\partial t} G(t, \tau)e(\tau)\, d\tau - G(t, t^+)e(t).$$

Since, as noted above, $G(t, \tau)$ is continuous, $G(t, t^-) = G(t, t^+)$. The second derivative follows in the same way,

$$x''(t) = \int_{t_0}^{t} \frac{\partial^2}{\partial t^2} G(t, \tau)e(\tau)\, d\tau + \frac{\partial}{\partial t} G(t, t^-)e(t)$$

$$+ \int_{t}^{t_1} \frac{\partial^2}{\partial t^2} G(t, \tau)e(\tau)\, d\tau - \frac{\partial}{\partial t} G(t, t^+)e(t)$$

$$= \int_{t_0}^{t_1} \frac{\partial^2}{\partial t^2} G(t, \tau)e(\tau)\, d\tau + e(t),$$

since, as noted above,

$$\frac{\partial G(t, t^-)}{\partial t} - \frac{\partial G(t, t^+)}{\partial t} = 1.$$

Then

$$L[x](t) = x''(t) + a(t)x'(t) + b(t)x(t)$$

$$= \int_{t_0}^{t_1} \left[\frac{\partial^2}{\partial t^2} G(t, \tau) + a(t)\frac{\partial}{\partial t} G(t, \tau) + b(t)G(t, \tau) \right] e(\tau)\, d\tau + e(t)$$

$$= \int_{t_0}^{t_1} L[G(t, \tau)]e(\tau)\, d\tau + e(t).$$

Since for fixed τ, $G(t, \tau)$ is also a solution of the unforced equation, $L[G(t, \tau)] = 0$. Hence,

$$L[x] = e,$$

and the existence part of the theorem is established.

Suppose that there are two solutions $x_1(t)$, $x_2(t)$ of the boundary value problem and let $\omega(t) = x_1(t) - x_2(t)$. Then $\omega(t_0) = x_1(t_0) - x_2(t_0) = 0$ and, similarly, $\omega(t_1) = 0$. Hence $\omega(t)$ is a solution of (7.1) that has a zero at t_0 and a zero at t_1, in violation of the hypothesis. This completes the proof.

To treat the general boundary conditions (2.9), (2.10), we use the idea of *superposition.* Let $\chi(t)$ be the unique solution of (2.12)–(2.14) given by Theorem 2.2 and let $\Omega(t)$ be the unique solution of (2.4), (2.2), (2.3) given by Theorem 2.1. Then $\Gamma(t) = \Omega(t) + \chi(t)$ is the desired solution of (2.8)–(2.10), since

$$L(\Gamma) = L(\Omega + \chi) = L(\Omega) + L(\chi) = 0 + e(t) = e(t),$$

so $\Gamma(t)$ satisfies (2.8). Further,

$$\Gamma(t_0) = \Omega(t_0) + \chi(t_0) = A + 0 = A$$
$$\Gamma(t_1) = \Omega(t_1) + \chi(t_1) = B + 0 = B$$

and the boundary conditions (2.9), (2.10) are satisfied.

Many other types of boundary conditions are of interest. We single out, in particular,

$$y(t_0) = A, \qquad y'(t_1) = m$$

and "periodic" conditions

$$y(t_0) = y(t_1), \qquad y'(t_0) = y'(t_1).$$

The general linear two-point boundary value problem with "separated conditions" takes the form

$$a_1 x(t_0) + a_2 x'(t_0) = A$$
$$b_1 x(t_1) + b_2 x'(t_1) = B.$$

We will not consider these problems here.

To illustrate the preceding theorems, we consider the following boundary value problem,

$$y'' + y = t \tag{2.15}$$

$$y(0) = 0 \tag{2.16}$$

$$y\left(\frac{\pi}{2}\right) = 1. \tag{2.17}$$

Now $\varphi(t) = \sin(t)$ and $\psi(t) = \cos(t)$ are linearly independent solutions of

$$y'' + y = 0, \tag{2.18}$$

and $\varphi(0) = 0$, $\psi(\pi/2) = 0$. Further, $W(\varphi, \psi)(t) = -1$. Hence any solution $x(t)$ of (2.18) can be written

$$x(t) = c_1 \cos(t) + c_2 \sin(t).$$

If $x(0) = 0$, then $c_1 = 0$, so any solution that has a zero at $t = 0$ is a multiple of $\sin(t)$ and cannot have a zero at $t = \pi/2$. Thus the basic hypothesis is satisfied.
Now with this choice of $\varphi(t)$, $\psi(t)$, (2.11) yields

$$G(t, \tau) = \begin{cases} -\sin(t)\cos(\tau) & 0 \le t \le \tau \le \dfrac{\pi}{2} \\[2mm] -\cos(t)\sin(\tau) & 0 \le \tau \le t \le \dfrac{\pi}{2}, \end{cases}$$

and a solution of (2.15) that is zero at the two endpoints is given by

$$x(t) = \int_0^{\pi/2} G(t, \tau)\tau \, d\tau.$$

To evaluate the right-hand integral, we split it into two separate integrals,

$$x(t) = -\int_0^t \tau \cos(t)\sin(\tau) \, d\tau - \int_t^{\pi/2} \tau \cos(\tau)\sin(t) \, d\tau$$

$$= -\cos(t)\int_0^t \tau \sin(\tau) \, d\tau - \sin(t)\int_t^{\pi/2} \tau \cos(\tau) \, d\tau$$

$$= -\cos(t)[\sin(\tau) - \tau \cos(\tau)]_{\tau=0}^{\tau=t} - \sin(t)[\cos(\tau) + \tau \sin(\tau)]_{\tau=t}^{\tau=\pi/2}$$

$$= -\cos(t)\sin(t) + t\cos^2(t) - \frac{\pi}{2}\sin(t) + \sin(t)\cos(t) + t\sin^2(t)$$

or

$$x(t) = t - \frac{\pi}{2}\sin(t).$$

Now take $\Omega(t) = \sin(t)$, which satisfies (2.16)–(2.18), and the solution of (2.15)–(2.17) is

$$\Gamma(t) = \sin(t) + t - \frac{\pi}{2}\sin(t)$$

$$= (\sin(t))\left(1 - \frac{\pi}{2}\right) + t.$$

The foregoing analysis is extremely important for theoretical purposes, but when the general solution to the differential equation can be found explicitly, a direct approach frequently involves less labor. For example, in the problem just considered, the general solution of

$$y'' + y = 0$$

is given by $y(t) = c_1 \cos(t) + c_2 \sin(t)$ and a particular solution of

$$y'' + y = t$$

can be seen (or computed, by the method of variation of parameters or the method of undetermined coefficients) to be $y_p(t) = t$. Thus all solutions of this differential equation are of the form

$$y(t) = c_1 \cos(t) + c_2 \sin(t) + t.$$

The condition $y(0) = 0$ becomes

$$c_1 = 0$$

and $y(\pi/2) = 1$ becomes

$$c_2 + \frac{\pi}{2} = 1$$

or

$$c_2 = 1 - \frac{\pi}{2}.$$

The solution of (2.15)–(2.17) is

$$y(t) = \left(1 - \frac{\pi}{2}\right) \sin(t) + t,$$

as expected.

We will ignore this "easy way" and illustrate the construction of the Green's function with another example. Consider the boundary value problem

$$y'' - y = \cos(t) \tag{2.19}$$

$$y(0) = 1 \tag{2.20}$$

$$y(1) = 0. \tag{2.21}$$

Take, as solutions of $y'' - y = 0$, $\sinh(t)$ and $\sinh(t-1)$. These are linearly independent solutions, each of which is zero at exactly one of the boundary points. The Wronskian is constant and at $t = 0$ takes the value $\sinh(1)$. The Green's function then takes the form

$$G(t,\tau) = \begin{cases} \dfrac{\sinh(t)\sinh(\tau-1)}{\sinh(1)}, & 0 \le t \le \tau \\ \dfrac{\sinh(\tau)\sinh(t-1)}{\sinh(1)}, & \tau \le t \le 1. \end{cases}$$

Thus the solution to the differential equation that is zero at $t = 0$ and at $t = 1$ is given by

$$x(t) = \int_0^1 G(t,\tau)\cos(\tau)\,d\tau.$$

To evaluate this integral it is easier to break it into two parts and compute

$$x(t) = \int_0^t G(t,\tau)\cos(\tau)\,d\tau + \int_t^1 G(t,\tau)\cos(\tau)\,d\tau.$$

If we define

$$f(t) = \int_t^1 \sinh(\tau-1)\cos(\tau)\,d\tau$$

$$= \frac{1}{2}[\cos(1) - \cos(t)\cosh(t-1) - \sinh(t)\sin(t-1)]$$

and

$$g(t) = \int_0^t \sinh(\tau)\cos(\tau)\,d\tau$$

$$= \frac{1}{2}[\cos(t)\cosh(t) + \sin(t)\sinh(t) - 1],$$

then

$$x(t) = \left(\frac{1}{\sinh(1)}\right)[\sinh(t-1)g(t) + \sinh(t)f(t)]$$

$$= \left(\frac{1}{2\sinh(1)}\right)[-\cos(t)\cosh(t-1)\sinh(t)$$

$$+ \cos(t)\cosh(t)\sinh(t-1) + \cos(1)\sinh(t) - \sinh(t-1)].$$

Making use of the identities

$$\sinh(x + y) = \sinh(x)\cosh(y) + \cosh(x)\sinh(y)$$
$$\cosh(x + y) = \cosh(x)\cosh(y) + \sinh(x)\sinh(y)$$

simplifies this to

$$x(t) = \left(\frac{1}{2\sinh(1)}\right)[-\cos(t)\sinh(1) + \cos(1)\sinh(t)$$
$$+ \cosh(t)\sinh(1) - \cosh(1)\sinh(t)].$$

This, of course, is the solution to the differential equation with zero boundary conditions. To complete the problem, we add to $x(t)$ a solution of $y'' - y = 0$ that satisfies $y(0) = 1$ and $y(1) = 0$. Such a solution must solve

$$c_1 = 1$$
$$c_1\cosh(1) + c_2\sinh(1) = 0.$$

Thus a complete solution of (2.19)–(2.21) is

$$y(t) = \cosh(t) + \left(-\frac{\cosh(1)}{\sinh(1)}\right)\sinh(t) + x(t).$$

EXERCISES

1. Use Green's functions to find solutions of the following boundary value problems.

(a) $y'' + y = t$
$y(0) = 0$
$y(1) = 1$

(b) $y'' + y = t^2$
$y(0) = 0$
$y\left(\frac{\pi}{2}\right) = 1$

(c) $y'' - y = 0$
$y(0) = 0$
$y(1) = 1$

(d) $y'' - y = t$
$y(0) = 0$
$y(1) = 1$

(e) $y'' + 2y' + y = t$
$y(0) = 0$
$y(1) = 1$

(f) $y'' - 3y' - 4y = t^2$
$y(0) = 0$
$y(1) = 2$

2. Consider the boundary value problem

$$L(y) = y'' + a(t)y' + b(t)y = 0$$

$$y(t_0) = A, \qquad y'(t_1) = m.$$

Prove a theorem corresponding to Theorem 2.1.

3. Find solutions of the following boundary value problems by finding the general solution of the differential equation and determining appropriate constants.

(a) $y'' + y = t$

$y(0) = 0$

$y'\left(\dfrac{\pi}{2}\right) = 0$

(b) $y'' - y = t$

$y(0) = 0$

$y'\left(\dfrac{1}{2}\right) = 0$

4. In Chapter 1 we noted that the system

$$z = \begin{bmatrix} 0 & 1 \\ -a(t) & -b(t) \end{bmatrix} z \qquad\qquad (*)$$

and the scalar equation (**) $y'' + a(t)y' + b(t)y = 0$ are equivalent. Let $\varphi(t)$ and $\psi(t)$ be solutions of (**). Show that $\begin{pmatrix} \varphi(t) & \psi(t) \\ \varphi'(t) & \psi'(t) \end{pmatrix}$ is a fundamental matrix for (*) on an interval I, if and only if the Wronskian of $\varphi(t)$, $\psi(t)$ is not zero on I.

5. Show that if $\varphi(t)$ and $\psi(t)$ are solutions of (2.4) with $\varphi(a) = \psi(a) = 0$, then $\varphi(t) = c\psi(t)$ for some constant c. (*Hint:* Compute the Wronskian.)

6. Use Exercise 5 to show that $\varphi(t)$ in (2.11) may be replaced by any other solution with a zero at t_0.

7. Use Theorem 3.5 of Chapter 1 to show that the Wronskian for two solutions of $y'' + a(t)y = 0$ is constant.

8. Show that the Wronskian of two solutions of $y'' + a(t)y' + b(t)y = 0$ is always zero or is never zero. (A proof can be made based on the corresponding system, as in Exercise 6. Alternatively, show directly that the Wronskian is a solution of $w' + a(t)w = 0$.)

3. Oscillation and Comparison Theorems

In this section we collect some facts that will be of value in the analysis that follows. First we note two alternate forms of the linear second-order equation

$$y'' + a(t)y' + b(t)y = 0. \qquad\qquad (3.1)$$

If $\varphi(t)$ is a solution of (3.1), then $\varphi(t)$ is also a solution of

$$(r(t)y')' + q(t)y = 0, \qquad\qquad (3.2)$$

where $r(t) = e^{\int_0^t a(s)\,ds}$ and $q(t) = b(t)e^{\int_0^t a(s)\,ds}$. Equation (3.2) is known as the "self-adjoint" form of (3.1). The next transformation is somewhat more complicated.

Let $a(t)$ be differentiable, and define a new dependent variable $u(t)$ by

$$u(t) = y(t)e^{1/2 \int_0^t a(s)\, ds},$$ (3.3)

where $y(t)$ is a solution of (3.1). Then

$$y'(t) = u'(t)e^{-1/2 \int_0^t a(s)\, ds} - \tfrac{1}{2}u(t)a(t)e^{-1/2 \int_0^t a(s)\, ds}$$

and

$$y''(t) = u''(t)e^{-1/2 \int_0^t a(s)\, ds} - u'(t)a(t)e^{-1/2 \int_0^t a(s)\, ds}$$
$$- \tfrac{1}{2}u(t)a'(t)e^{-1/2 \int_0^t a(s)\, ds} + \tfrac{1}{4}u(t)a^2(t)e^{-1/2 \int_0^t a(s)\, ds}.$$

Substituting into (3.1), after dividing out $e^{-1/2 \int_0^t a(s)\, ds}$, yields

$$u''(t) + \left(b(t) - \frac{1}{2}a'(t) - \frac{a^2}{4}\right)u(t) = 0.$$ (3.4)

Note that, from (3.3), $u(t)$ and $y(t)$ have the same zeros, so that for questions involving the location of zeros, equation (3.4) may be used.

Equation (3.2) may be simplified by the change of independent variable

$$\tau = \int_0^t \frac{ds}{r(s)},$$

where we assume

$$\lim_{t \to \infty} \int_0^t \frac{ds}{r(s)} = \infty.$$

A straightforward calculation yields

$$\frac{d^2 y}{d\tau^2} + r(\tau)q(\tau)y = 0.$$ (3.5)

Both (3.4) and (3.5) take the form

$$y'' + Qy = 0$$ (3.6)

and we confine our attention to equations of the form (3.6).

The first result is a simple comparison of solutions of a first-order linear inequality.

LEMMA 3.1

Let $\varphi(t)$ and $\psi(t)$ be differentiable functions satisfying

$$\varphi'(t) + a(t)\varphi(t) \leq b(t) \tag{3.7}$$

and

$$\psi'(t) + a(t)\psi(t) \geq b(t), \tag{3.8}$$

respectively, on an interval I. If $\varphi(t_0) = \psi(t_0)$, $t_0 \in I$, then

$$\varphi(t) \leq \psi(t), \qquad t \geq t_0, \qquad t \in I,$$

and

$$\varphi(t) \geq \psi(t), \qquad t \leq t_0, \qquad t \in I.$$

Further, if $\varphi(t_0) < \psi(t_0)$, then $\varphi(t) < \psi(t)$ for $t \geq t_0$ or if either inequality (3.7) or (3.8) is strict, then $\varphi(t) < \psi(t)$ for $t > t_0$.

Proof: The inequality

$$\varphi'(t) + a(t)\varphi(t) \leq \psi'(t) + a(t)\psi(t)$$

leads to

$$(\varphi(t) - \psi(t))' + a(t)(\varphi(t) - \psi(t)) \leq 0.$$

Multiplying both sides by $e^{\int_{t_0}^{t} a(s)\,ds}$ produces

$$[e^{\int_{t_0}^{t} a(s)\,ds}(\varphi(t) - \psi(t))]' \leq 0. \tag{3.9}$$

Since the function in square brackets is nonincreasing, its value at $t > t_0$ is less than or equal to its value at t_0, or

$$e^{\int_{t_0}^{t} a(s)\,ds}(\varphi(t) - \psi(t)) \leq \varphi(t_0) - \psi(t_0) = 0, \tag{3.10}$$

from which it follows that

$$\varphi(t) \leq \psi(t).$$

If $\varphi(t_0) < \psi(t_0)$, (3.10) yields

$$\varphi(t) < \psi(t).$$

If either inequality (3.7) or (3.8) is strict, then (3.9) becomes a strict inequality and the function in square brackets in (3.9) is strictly decreasing, so that (3.10) becomes

$$e^{\int_{t_0}^{t} a(s)\,ds}(\varphi(t) - \psi(t)) < \varphi(t_0) - \psi(t_0)$$

or

$$\varphi(t) < \psi(t), \qquad t > t_0.$$

As a consequence of this lemma, we can prove a comparison result for nonlinear equations that will be useful in the remainder of this chapter.

THEOREM 3.2   *(comparison theorem)*

Let $f(t, y)$ be a continuous function satisfying a Lipschitz condition with respect to y in a region G. Let $x(t)$ be a solution of

$$x' \le f(t, x) \tag{3.11}$$

and let $y(t)$ be a solution of

$$y' = f(t, y) \tag{3.12}$$

with $y(t_0) = x(t_0)$, $(t_0, y(t_0)) \in G$. Then $x(t) \le y(t)$ for $t > t_0$, as long as $(t, x(t))$ and $(t, y(t))$ remain in G. Further, if $x(t_0) < y(t_0)$, or if the inequality in (3.11) is strict, then $x(t) < y(t)$ for $t > t_0$, as long as $(t, x(t))$ and $(t, y(t))$ remain in G.

Proof: Suppose that the theorem is false; then there is a time $t_1 > t_0$ when $x(t_1) > y(t_1)$. Since the functions start out equal there is a last point, $t_2 < t_1$, when $x(t_2) = y(t_2)$ ($t_2 = t_0$ is not excluded), and hence $x(t) > y(t)$ on $(t_2, t_1]$. Let $\sigma(t) = x(t) - y(t)$. Then, for $t_2 < t \le t_1$, $\sigma(t) > 0$ and

$$
\begin{aligned}
\sigma'(t) &= x'(t) - y'(t) \\
&\le f(t, x(t)) - f(t, y(t)) \\
&\le |f(t, x(t)) - f(t, y(t))| \\
&\le L|x(t) - y(t)| \\
&= L(x(t) - y(t)), \quad \text{for} \quad t_2 < t < t_1.
\end{aligned}
$$

Hence,

$$\sigma'(t) \le L\sigma(t), \quad \text{for} \quad t_2 < t < t_1, \tag{3.13}$$

and

$$\sigma(t_2) = 0.$$

Use (3.13) and

$$z' = Lz$$

$$z(t_2) = 0$$

in Lemma 3.1 to conclude that

$$\sigma(t) \le z(t) = z(t_2)e^{L(t - t_2)} = 0.$$

Hence, $\sigma(t) \equiv 0$, $t \in [t_2, t_1]$, contradicting $\sigma(t_1) > 0$.

If $x(t_0) < y(t_0)$, let $t_1 > t_0$ be the first point when $x(t) = y(t)$, let $\sigma(t) = y(t) - x(t)$, $t_0 \leq t < t_1$, and carry out the same proof.

If the inequality in (3.11) is strict, then

$$\sigma'(t) < L\sigma$$

and Lemma 3.1 also applies.

We use the comparison theorem and the polar coordinate technique introduced in Chapter 2 to prove a classic theorem.

THEOREM 3.3 *(Sturm separation theorem)*
Let $\varphi(t)$ and $\psi(t)$ be solutions of

$$y'' + Q_1(t)y = 0 \tag{3.14}$$

and

$$x'' + Q_2(t)x = 0, \tag{3.15}$$

respectively. If $Q_1(t) \geq Q_2(t)$, then $\varphi(t)$ has a zero between (in the wide sense) any two consecutive zeros of $\psi(t)$.

Proof: First convert (3.14) and (3.15) to systems of the form

$$\begin{aligned} y_1' &= y_2 \\ y_2' &= -Q_1(t)y_1 \end{aligned} \tag{3.16}$$

$$\begin{aligned} x_1' &= x_2 \\ x_2' &= -Q_2(t)x_1. \end{aligned} \tag{3.17}$$

Introduce the polar coordinate transformations $x_1 = r\cos(\theta)$, $x_2 = r\sin(\theta)$, and $y_1 = \rho\cos(\omega)$, $y_2 = \rho\sin(\omega)$. Then $\theta(t)$ and $\omega(t)$ are solutions of the differential equations (see equation (1.5) of Chapter 2)

$$\theta' = -\sin^2(\theta) - Q_2(t)\cos^2(\theta) \tag{3.18}$$

$$\omega' = -\sin^2(\omega) - Q_1(t)\cos^2(\omega). \tag{3.19}$$

However, since $Q_1(t) \geq Q_2(t)$, $\omega(t)$ also satisfies

$$\omega' \leq -\sin^2(\omega) - Q_2(t)\cos^2(\omega). \tag{3.20}$$

Since $\sin^2(\omega)$ and $\cos^2(\omega)$ are continuously differentiable, the right-hand side of (3.18) is Lipschitzian, so Theorem 3.2 applies to equation (3.18) and inequality (3.20) to allow us to conclude that if $\omega(t_0) \leq \theta(t_0)$ for any t_0, then

$$\omega(t) \leq \theta(t), \qquad t \geq t_0. \tag{3.21}$$

Let t_1 and t_2 be the two consecutive zeros of $\psi(t)$ given in the hypotheses of the theorem. Without loss of generality we can assume that $\psi'(t_1) > 0$, $\psi'(t_2) < 0$ (since $-\psi(t)$ is also a solution with zeros at t_1 and at t_2). Zeros of $\psi(t)$ correspond to places where $\theta(t)$ passes through $n\pi + \pi/2$. We can assume that $\theta(t_1) = \pi/2$ and $\theta(t_2) = -\pi/2$ without loss of generality. If $\varphi(t)$ does not have a zero on $[t_1, t_2]$, it is of one sign there, which, again, without loss of generality we take to be positive. Thus $\omega(t)$ satisfies

$$-\frac{\pi}{2} < \omega(t) < \frac{\pi}{2}, \qquad t \in [t_1, t_2].$$

(The phase plane trajectory lies in the right half-plane.) Therefore, since $\theta(t_1) = \pi/2 > \omega(t_1)$, (3.21) requires

$$\omega(t_2) < \theta(t_2) = -\frac{\pi}{2}.$$

Thus there must be a value $t_3 < t_2$ such that $\omega(t_3) = -\pi/2$ or $\varphi(t_3) = 0$, which is the desired contradiction. The supposition that $\varphi(t) > 0$ on $[t_1, t_2]$ is false and the theorem is established.

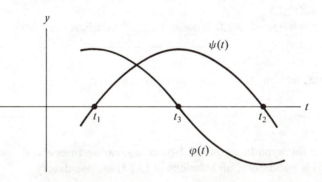

Figure 3.1. An illustration of the Sturm comparison theorem.

Figure 3.1 illustrates this point. Note that if $Q_1(t) \equiv Q_2(t)$, then the theorem applies to both $\varphi(t)$ and $\psi(t)$. More specifically, we have the following.

COROLLARY 3.4

The zeros of linearly independent solutions of

$$y'' + A(t)y = 0 \tag{3.22}$$

interlace.

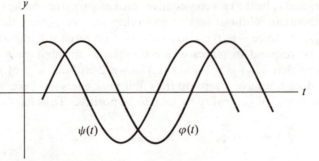

Figure 3.2 The interlacing of zeros of linearly independent solutions of second-order equations.

Corollary 3.4 is illustrated in Figure 3.2. A solution $\varphi(t)$ of (3.22) is said to be *oscillatory* if there is a sequence t_n, $\lim_{n\to\infty} t_n = \infty$, such that $\varphi(t_n) = 0$. A consequence of Corollary 3.4 is the following.

COROLLARY 3.5
   If one solution of (3.22) is oscillatory, all solutions are oscillatory.

In fact, we can state a more general corollary.

COROLLARY 3.6
   Under the hypotheses of the Sturm separation theorem, if one solution of (3.15) is oscillatory, all solutions of (3.14) are oscillatory.

   The Sturm separation theorem is frequently used to provide an oscillation criterion for a variable coefficient equation that can be compared with a constant coefficient equation. For example, consider

$$y'' + Q(t)y = 0 \tag{3.23}$$

where $Q(t) \geq \varepsilon > 0$, for $t > t_0$. Then solutions of (3.23) can be compared to solutions of

$$y'' + \varepsilon y = 0.$$

This last equation has a solution $\cos(\sqrt{\varepsilon})t$, which is oscillatory, so all solutions of (3.23) must be oscillatory. All solutions of

$$y'' + t^n y = 0$$

must be oscillatory, $n \geq 1$, by this comparison technique with $\varepsilon = 1$ and $t_0 = 1$.

EXERCISES

1. Show that $y'' + a(t)y = 0$ is oscillatory if $a(t) > (1 + \varepsilon)/4t^2$, $t \geq 1$, $\varepsilon > 0$, and that it is not oscillatory if $a(t) < (1 - \varepsilon)/4t^2$, $t \geq 1$, $\varepsilon > 0$. (*Hint:* Compare with the Euler equation $y'' + (1/4t^2)y = 0$.)

2. Show that no solution of $y'' + a(t)y = 0$ is oscillatory if $a(t) < 0$, $t \geq t_0$.

3. Use Lemma 3.1 to find upper and lower bounds for a solution of $y' + y = \sin(y)$, $y(0) = 1$.

4. Use the equation for the polar radius (equation (5.6) in Chapter 2) to find region G such that for t sufficiently large, $(x(t), y(t)) \in G$, where $(x(t), y(t))$ is a solution of

$$x' = x + y - x(x^2 + y^2)$$
$$y' = -x + y - y(x^2 + y^2).$$

5. If a solution $\varphi(t)$ of $x'' + Q(t)x = 0$ is oscillatory, show that if $\theta(t)$ is the corresponding polar angle, then $\lim_{t \to \infty} |\theta(t)| = \infty$.

6. Show that the zeros of two linearly dependent, nontrivial solutions of (3.22) coincide.

4. Sturm-Liouville Problems

We begin our study with a simple example. Consider the boundary value problem

$$y'' + \lambda^2 y = 0 \qquad\qquad (4.1)$$

$$y(0) = y(1) = 0 \qquad\qquad (4.2)$$

where λ is a real number. There is always the solution $y(t) \equiv 0$, but we seek a nontrivial solution. The equation (4.1) can be solved directly; all solutions take the form $y(t) = c_1 \cos(\lambda t) + c_2 \sin(\lambda t)$. For this solution to be zero at $t = 0$, it must be the case that $c_1 = 0$. Thus any nontrivial solution takes the form

$y(t) = c_2 \sin(\lambda t)$. To satisfy the second condition in (4.2) requires

$$y(1) = c_2 \sin(\lambda) = 0.$$

The choice $c_2 = 0$ produces the trivial solution, so to have a nontrivial solution it must be the case that $\sin(\lambda) = 0$, or

$$\lambda = n\pi, \qquad n = 0, \pm 1, \pm 2, \dots.$$

For values $\lambda = n\pi$, there will be a one-parameter family of solutions

$$y(t) = c_2 \sin(n\pi t)$$

and for $\lambda \neq n\pi$, there is only the trivial solution. These special values of λ for which there are nontrivial solutions are called *eigenvalues* and the corresponding solutions are called *eigenfunctions*.

Problems of the form

$$(r(t)y'(t))' + [\lambda p(t) - q(t)]y(t) = 0$$
$$\alpha_1 y(a) + \alpha_2 y'(a) = 0$$
$$\beta_1 y(b) + \beta_2 y'(b) = 0$$

are called *Sturm-Liouville problems*. They occur in a variety of applications after the separation of variables in problems in partial differential equations. We seek to find the special values of λ that yield nontrivial solutions, to find those solutions, and to identify properties of such solutions. In this chapter the special case $r \equiv 1$ is treated, although the problems will often indicate changes for the case of a variable $r(t)$. There are also changes of variables that make $r \equiv 1$, as shown in Section 3.

EXERCISES

1. Find the eigenvalues for the following.

   (a) $y'' + \lambda y = 0$
        $y(0) = 0, \qquad y'(1) = 0$

   (b) $y'' + \lambda y = 0$
        $y'(0) = 0, \qquad y(1) = 0$

   (c) $y'' + \lambda y = 0$
        $y(1) = 0, \qquad y'(1) = 0$

2. Find the eigenfunctions $y(t, \lambda)$, corresponding to the eigenvalues in Exercise 1, that satisfy $\int_0^1 y^2(t, \lambda) \, dt = 1$.

5. The Existence of Eigenvalues for Sturm-Liouville Problems

Consider the Sturm-Liouville problem

$$y'' + [\lambda p(t) - q(t)]y = 0 \tag{5.1}$$

$$\alpha_1 y(a) + \alpha_2 y'(a) = 0 \tag{5.2}$$

$$\beta_1 y(b) + \beta_2 y'(b) = 0 \tag{5.3}$$

with $p(t)$ and $q(t)$ continuous, $p(t) > 0$ on $[a, b]$, and $\alpha_1 \beta_1 > 0$. As noted in the preceding section, values of λ for which there is a nontrivial solution of (5.1)–(5.3) are called *eigenvalues* and a nontrivial solution of (5.1)–(5.3) associated with an eigenvalue is called an *eigenfunction*. Note that since the equation (5.1) and the boundary conditions (5.2), (5.3) are linear, if $\varphi(t)$ is an eigenfunction, so is $c\varphi(t)$ where c is a constant. The set of all eigenvalues is called the *spectrum* of (5.1)–(5.3).

If the equation is converted to a system of the form

$$y_1' = y_2$$
$$y_2' = [q(t) - \lambda p(t)]y_1 \tag{5.4}$$

$$\ell_1 : \alpha_1 y_1(a) + \alpha_2 y_2(a) = 0$$
$$\ell_2 : \beta_1 y_1(b) + \beta_2 y_2(b) = 0, \tag{5.5}$$

we get a simple geometric preview of the problem. In the $y_1 - y_2$ plane, the boundary conditions can be viewed as lines through the origin—see Figure 5.1.

We seek a solution of the system of differential equations (5.4) that starts at $t = a$ on the line $\ell_1 : \alpha_1 y_1 + \alpha_2 y_2 = 0$ and that at $t = b$ lies on the line $\ell_2 : \beta_1 y_1 + \beta_2 y_2 = 0$. Of course, the solution may "circle" the origin many times before terminating on ℓ_2 at time $t = b$. If λ is not an eigenvalue, by definition the only solution that starts on ℓ_1 at $t = a$ and terminates on ℓ_2 at $t = b$ is the trivial solution, represented by the origin in Figure 5.1.

This view of the boundary value problem is called a "shooting method" when we attempt to realize it in a numerical computation. The "idea" is to find the "correct" initial value problem to satisfy the boundary conditions. The first boundary condition prescribes that the solution must "start" on the line ℓ_1. The object becomes selecting the right initial conditions on this line—the right starting point—so that the solution starting at this point will terminate on the second line, ℓ_2, and thus satisfy the second boundary condition. This idea is not unlike that of the solution map considered in Chapter 2 for autonomous systems.

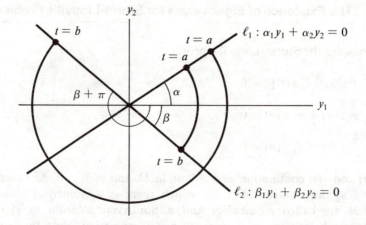

Figure 5.1 A geometric view of linear, two-point boundary conditions.

Indeed, if the differential equation should be nonlinear and autonomous, those very ideas are useful. We want to look at the set of initial value problems satisfying the first boundary condition (lying on ℓ_1 at $t = a$) and ask where these solutions are at $t = b$. Perhaps one (or more) will lie on the line ℓ_2. Because the differential equation is linear, the solutions will "move" (as λ varies) in a very regular way and either all or none will satisfy this condition. (For nonlinear equations the same ideas are useful, even though the behavior of the solutions will not be so regular.) The idea, then, is to vary λ until all of the elements of the set of solutions that start at $t = a$ on ℓ_1 terminate at $t = b$ on ℓ_2. This value of λ will be an eigenvalue.

If the system (5.4) is converted to polar coordinates, this can be expressed in an even nicer way. Following the procedure of Chapter 2, the solutions $(y_1(t), y_2(t))$ of the system above can be viewed as the solutions $(r(t), \theta(t))$ of

$$r'(t) = r[1 + q - \lambda p] \sin(\theta) \cos(\theta)$$

$$\theta'(t) = -\sin^2(\theta) - (\lambda p - q) \cos^2(\theta). \tag{5.6}$$

To emphasize the dependence of solutions on the parameter λ, denote solutions by $r(t, \lambda)$ and $\theta(t, \lambda)$. The polar angle $\theta(t, \lambda)$ allows us to state the boundary conditions in terms of the angles α and β, the slopes of the lines ℓ_1 and ℓ_2 ($\tan(\alpha) = -\alpha_1/\alpha_2$, $\tan(\beta) = -\beta_1/\beta_2$). Specifically, we seek a value of the parameter λ so that

$$\theta(a, \lambda) = \alpha$$

$$\theta(b, \lambda) = \beta - n\pi. \tag{5.7}$$

Since the differential equation for θ is independent of r, all solutions of (5.6) with a fixed initial angle, $\theta(a, \lambda) = \alpha$, "move" around the origin together in the sense that $\theta(t, \lambda)$ is the same, independent of the initial value $r(a)$. Thus, solving (5.4), (5.5) can be viewed as a "hunting" procedure in λ—fix (i.e., guess) λ, solve the initial value problem

$$\theta'(t, \lambda) = -\sin^2(\theta)(t, \lambda) - (\lambda p(t) - q(t)) \cos^2(\theta(t, \lambda))$$
$$\theta(a, \lambda) = \alpha, \tag{5.8}$$

observe $\theta(b, \lambda)$ to see if $\theta(b, \lambda) < \beta$ or $\theta(b, \lambda) > \beta$, correct the initial guess λ, accordingly, and repeat. (As noted previously, this is the essence of a "shooting method.") The task would be easier if we had a hint as to how $\theta(b, \lambda)$ changes as λ changes. Such guidance is given in the following lemma.

LEMMA 5.1

For a fixed $T > a$, the solution of (5.8), $\theta(T, \lambda)$, is a strictly decreasing function of λ.

In particular, $\theta(b, \lambda)$ is decreasing in λ, so that if for a trial value, $\lambda = \lambda_1$, we find $\theta(b, \lambda_1) < \beta$, the new "guess," $\lambda = \lambda_2$, should satisfy $\lambda_2 < \lambda_1$.

Proof of Lemma 5.1:   Let $\lambda_1 > \lambda_2$ be two values of the parameter and let $\theta(t, \lambda_1)$, $\theta(t, \lambda_2)$ be the corresponding solutions of (5.8). Then

$$\theta'(t, \lambda_1) = -\sin^2(\theta)(t, \lambda_1) - (\lambda_1 p - q) \cos^2(\theta)(t, \lambda_1)$$
$$\leq -\sin^2(\theta)(t, \lambda_1) - (\lambda_2 p - q) \cos^2(\theta)(t, \lambda_1),$$

since $-\lambda_1 p \cos^2(\theta)(t, \lambda_1) \leq -\lambda_2 p \cos^2(\theta)(t, \lambda_1)$, where $\lambda_1 > \lambda_2$, $p > 0$, and $\cos^2(\theta) \geq 0$. Thus

$$\theta'(t, \lambda_1) \leq -\sin^2(\theta)(t, \lambda_1) - (\lambda_2 p(t) - q(t)) \cos^2(\theta)(t, \lambda_1)$$
$$\theta'(t, \lambda_2) = -\sin^2(\theta)(t, \lambda_2) - (\lambda_2 p(t) - q(t)) \cos^2(\theta)(t, \lambda_2)$$

and we are in exactly the situation described in Theorem 3.2 (the comparison theorem). Thus for any fixed $T > a$,

$$\theta(T, \lambda_1) \leq \theta(T, \lambda_2).$$

In particular, this is true at $T = b$. It remains to be shown that the inequality is strict.

Without loss of generality, we can assume that $-\pi/2 < \alpha \le \pi/2$. If $\alpha \ne \pi/2$, then $\theta'(a, \lambda_2) > \theta'(a, \lambda_1)$, so for some interval $(a, t_0]$, $\theta(t, \lambda_1) < \theta(t, \lambda_2)$. Now, from the second part of Theorem 3.2 it follows that $\theta(t, \lambda_1) < \theta(t, \lambda_2)$ for $t \in [t_0, b]$, and hence for $t \in (a, b]$, as claimed in the theorem.

If $\alpha = \pi/2$, the problem is somewhat more delicate. Since $\theta(t, \lambda_1) \le \theta(t, \lambda_2)$, if there is no t_0 such that $\theta(t_0, \lambda_1) < \theta(t_0, \lambda_2)$ (in which case the argument above would apply), $\theta(t, \lambda_1) = \theta(t, \lambda_2)$, $t \in [a, b]$. Further, $\theta'(a, \lambda_i) < 0$, so $\theta(t, \lambda_i)$ is not constant. Then for every $\varepsilon > 0$, there exists an angle $\alpha^* \ne \pi/2$ and a time $t^* \in (a, a + \varepsilon)$ such that $\theta'(t^*, \lambda_1) < \theta'(t^*, \lambda_2)$, since $\cos(\alpha^*) \ne 0$. Then $\theta(t, \lambda_1) < \theta(t, \lambda_2)$ just to the right of t^*, contradicting $\theta(t, \lambda_1) = \theta(t, \lambda_2)$, $a \le t \le b$, in turn contradicting the nonexistence of t_0 above. Thus $\theta(b, \lambda)$ is *strictly* decreasing in λ.

The principal result of this section may now be stated.

THEOREM 5.2

There exists an infinite number of eigenvalues $\lambda_0 < \lambda_1 < \lambda_2 < \cdots$ for (5.1)–(5.3).

The proof proceeds by showing that, as a function of λ, $\theta(b, \lambda)$ takes on all of the values $\beta - n\pi$, $n = 0, 1, \dots$. The argument will depend on the zeros of a certain comparison equation, for which we need the following property of $\theta(t, \lambda)$.

LEMMA 5.3

If $\theta(t_n, \lambda) = \pi/2 - n\pi$, $\theta(t, \lambda) < \pi/2 - n\pi$ for $t > t_n$.

Proof: At a point t where $\theta(t, \lambda) = \pi/2 - n\pi$,

$$\theta'(t, \lambda) = -1 < 0.$$

Once $\theta(t, \lambda)$ crosses the value $\theta = \pi/2 - n\pi$, it can never recross it. The solution is illustrated in Figure 5.2.

As a consequence of this lemma, if a solution of (5.1) has k zeros on an interval $[a, T]$, then

$$\theta(T, \lambda) < \theta(a, \lambda) - (k - 1)\pi.$$

This property will be used to prove the following.

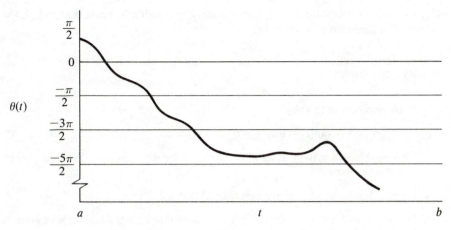

Figure 5.2 A plot of $\theta(t\lambda)$ as a function of t.

LEMMA 5.4

$\mathrm{Lim}_{\lambda \to \infty}\, \theta(T, \lambda) = -\infty$ for every $T \in (a, b]$.

Proof: To prove the lemma, we seek to compare it with a solution of an equation that can be solved explicitly—an equation with constant coefficients. Since $p(t)$ and $q(t)$ are continuous and $p(t) > 0$ on the closed interval $[a, b]$, there is a positive constant P such that $0 < P < p(t)$. Similarly, there is a constant Q such that $Q \geq q(t)$, $t \in [a, b]$. Let $x(t)$ be a solution of

$$x'' + (\lambda P - Q)x = 0 \qquad (5.9)$$

and denote its polar angle by $\psi(t, \lambda)$. The corresponding differential equation for ψ is

$$\psi' = -\sin^2(\psi) + (Q - \lambda P)\cos^2(\psi). \qquad (5.10)$$

It follows that, for any $T \in (a, b]$,

$$\theta(T, \lambda) < \psi(T, \lambda), \qquad (5.11)$$

since $\lambda P - Q \leq \lambda p(t) - q(t)$. However, solutions of (5.9), for $\lambda P - Q > 0$, have zeros spaced at intervals of length $[\lambda P - Q]^{-1/2}\pi$. As $\lambda \to \infty$, the distance between zeros tends to zero, and hence the number of zeros in $[a, b]$ becomes unbounded. By the remark following the previous lemma, $\lim_{\lambda \to \infty} \psi(T, \lambda) = -\infty$. Thus from (5.11), $\lim_{\lambda \to \infty} \theta(T, \lambda) = -\infty$.

As $\lambda \to \infty$, $\theta(b, \lambda) = \beta - n\pi$ for every n. Label λ_n as the value where $\theta(b, \lambda_n) = \beta - n\pi$. Any solution satisfying $\theta(a, \lambda_n) = \alpha$ is an eigenfunction. Since

$\theta(b, \lambda)$ is strictly monotone in λ, there is only one value λ_n such that $\theta(b, \lambda_n) = \beta - n\pi$. The theorem now follows.

EXERCISES

1. Replace equation (5.1) by

$$(r(t)y')' + (\lambda p(t) - q(t))y = 0, \qquad r(t) > 0. \tag{*}$$

Change this to a system using $y_1' = y_2/r(t)$ as the first equation. Prove the corresponding Lemma 5.1.

2. Using (*) of Exercise 1, prove Lemmas 5.3 and 5.4.

3. If $-\pi/2 < \beta < 0 < \alpha < \pi/2$ and $q(t) > 0$, show that $\lambda_0 > 0$ where λ_0 is as given in Theorem 5.1.

4. What if $\alpha < 0 < \beta$ in Exercise 3?

5. Suppose that $\alpha_1 \alpha_2 \neq 0$ in (5.1)–(5.3). Show that to each λ_n there cannot be two linearly independent eigenfunctions $\varphi_n(t, \lambda_n)$ and $\psi_n(t, \lambda_n)$. (*Hint:* Suppose the contrary and compute the Wronskian.)

6. In Exercise 5, can you remove the restriction $\alpha_1 \alpha_2 \neq 0$?

6. Two Properties of Eigenfunctions

The existence theorem of the previous section showed that there is a sequence of eigenvalues $\lambda_0 < \lambda_1 < \cdots < \lambda_n < \cdots$ such that if $\lambda = \lambda_i$, then there is a nontrivial solution of the Sturm-Liouville problem

$$y'' + [\lambda p(t) - q(t)]y = 0 \tag{6.1}$$

$$\alpha_1 y(a) + \alpha_2 y'(a) = 0 \tag{6.2}$$

$$\beta_1 y(b) + \beta_2 y'(b) = 0, \tag{6.3}$$

where $p(t) > 0$ and $\alpha_1 \beta_1 > 0$. Of course, if $\varphi(t, \lambda_i)$ is a solution, then any constant multiple is also an eigenfunction. From this family we can select a particular member, and we do this by choosing the eigenfunction for which

$$\int_a^b \frac{\varphi^2(t, \lambda_1)\, dt}{p(t)} = 1. \tag{6.4}$$

This is called the *normalized eigenfunction*. In this section we establish two properties of eigenfunctions that are important for applications.

Two functions $\varphi(t)$, $\psi(t)$ defined on an interval $[a, b]$ are said to be *orthogonal with respect to* the weight function $r(t) > 0$ if

$$\int_a^b r(t)\varphi(t)\psi(t)\,dt = 0. \tag{6.5}$$

When $r(t) \equiv 1$, the "with respect to ..." is dropped and $\varphi(t)$ and $\psi(t)$ are simply said to be *orthogonal*.

THEOREM 6.1

   Eigenfunctions of (6.1)–(6.3) that correspond to distinct eigenvalues are orthogonal on $[a, b]$ with respect to $p(t)$.

Proof: Let $\varphi_i(t)$ and $\varphi_j(t)$ be solutions of (6.1)–(6.3) that correspond to λ_i and λ_j, respectively, $\lambda_i \neq \lambda_j$; that is,

$$\varphi_i''(t) + (\lambda_i p(t) - q(t))\varphi_i(t) = 0$$

and

$$\varphi_j''(t) + (\lambda_j p(t) - q(t))\varphi_j(t) = 0.$$

Then, multiplying the first of the equations by $\varphi_j(t)$ and the second by $\varphi_i(t)$ yields

$$\varphi_j(t)\varphi_i''(t) + (\lambda_i p(t) - q(t))\varphi_i(t)\varphi_j(t) = 0$$

$$\varphi_i(t)\varphi_j''(t) + (\lambda_j p(t) - q(t))\varphi_i(t)\varphi_j(t) = 0.$$

Subtracting produces

$$\varphi_j(t)\varphi_i''(t) - \varphi_i(t)\varphi_j''(t) = (\lambda_j - \lambda_i)p(t)\varphi_i(t)\varphi_j(t)$$

or

$$(\varphi_j(t)\varphi_i'(t) - \varphi_i(t)\varphi_j'(t))' = (\lambda_j - \lambda_i)p(t)\varphi_i(t)\varphi_j(t). \tag{6.6}$$

If both sides of the inequality are integrated, on $[a, b]$ we have

$$(\lambda_j - \lambda_i)\int_a^b p(t)\varphi_i(t)\varphi_j(t)\,dt \tag{6.7}$$

$$= \varphi_j(b)\varphi_i'(b) - \varphi_i(b)\varphi_j'(b) - \varphi_j(a)\varphi_i'(a) + \varphi_i(a)\varphi_j'(a).$$

The theorem will be established if the right-hand side can be shown to be zero. Applying the boundary conditions (6.2) and (6.3) to the right-hand side of (6.7) yields

$$-\frac{\beta_2}{\beta_1}\varphi_j'(b)\varphi_i(b) + \frac{\beta_2}{\beta_1}\varphi_i'(b)\varphi_j(b) + \frac{\alpha_2}{\alpha_1}\varphi_j'(a)\varphi_i(a) - \frac{\alpha_2}{\alpha_1}\varphi_i'(a)\varphi_j(a) = 0,$$

which establishes the theorem.

The function $p(t)$, which multiplies the parameter in (6.1), is sometimes inconvenient in working with Sturm-Liouville problems and it is desirable to "scale it out." If $p(t)$ has sufficient derivatives, this can be done with the *Liouville transformation*, defined by

$$w(x) = p^{1/4}(x)y(x)$$

$$x(t) = \int_a^t p^{1/2}(s)\,ds \tag{6.8}$$

where the positive root is intended. Note that this is a change of both independent and dependent variables. Further, $dx/dt = p^{1/2}(t) > 0$. The computation is straightforward but very tedious. The end result is that w satisfies

$$w'' + [\lambda - Q(x)]w = 0 \tag{6.9}$$

where

$$Q(x) = \frac{-5}{16}p^{-3}(x)(p'(x))^2 + \frac{1}{4}p^{-2}(x)p''(x) + q(x)p^{-1}(x).$$

Since $p(x) > 0$, no singularities are introduced in (6.9), but, of course, $p''(x)$ must be assumed to be continuous if the basic existence theorems are to apply.

At $t = a$, $x = 0$, so the first boundary condition takes the form

$$\alpha_1 y(a) + \alpha_2 y'(a)$$
$$= \alpha_1 w(0)p^{-1/4}(0) + \alpha_2 w'(0)p^{-1/4}(0) - \tfrac{1}{4}\alpha_2 w(0)p^{-5/4}(0)p'(0)$$
$$= w(0)[\alpha p^{-1/4}(0) - \tfrac{1}{4}\alpha_2 p^{-5/4}(0)p'(0)] + \alpha_2 w'(0)p^{-1/4}(0) = 0.$$

If we define

$$\gamma_1 = \alpha_1 p^{-1/4}(0) - \frac{1}{4}\alpha_2 p^{-5/4}(0)p'(0)$$

$$\gamma_2 = \alpha_2 p^{-1/4}(0),$$

then the boundary condition takes the form

$$\gamma_1 w(0) + \gamma_2 w'(0) = 0.$$

Similarly, the other boundary condition may be rewritten at $x = c$, corresponding to

$$c = \int_a^b p^{1/2}(\xi)\, d\xi$$

as $\delta_1 w(c) + \delta_2 w'(c) = 0$. The new Sturm-Liouville problem then takes the form

$$w'' + (\lambda - Q(x))w = 0 \tag{6.10}$$

$$\gamma_1 w(0) + \gamma_2 w'(0) = 0 \tag{6.11}$$

$$\delta_1 w(c) + \delta_2 w'(c) = 0. \tag{6.12}$$

Hence, if $p''(\xi)$ is continuous and $p(\xi) > 0$, then the Sturm-Liouville problems (6.1)–(6.3) and (6.10)–(6.12) are equivalent.

Also, the weight function may be dropped in Theorem 6.1.

THEOREM 6.1'
   Eigenfunctions corresponding to distinct eigenvalues in the Sturm-Liouville problem (6.10–(6.12) are orthogonal.

The second property we wish to note is contained in the next theorem.

THEOREM 6.2
   An eigenfunction corresponding to the nth eigenvalue has exactly n zeros in $[a, b]$.

Discussion of a proof: The proof given for Theorem 5.1 in the preceding section—see particularly Figure 5.2—suggests that for each λ_n, $n = 0, 1, 2,$ ... given by this theorem, the corresponding solution has n zeros, since each value $\pi/2 - n\pi$ is crossed only once by the polar angle $\theta(t, \lambda_n)$. Perhaps, however, there are other eigenvalues and eigenfunctions for $\theta(t, \lambda)$ going "the other way," that is, λ such that $\theta(t, \lambda) = \beta + n\pi$. That this is not the case would follow from the next lemma.

LEMMA 6.3
   $\lim_{\lambda \to -\infty} \theta(t, \lambda) = \pi/2.$

The proof of this lemma is "tricky," and we omit it.

EXERCISES

1. Let $f(t)$, $g(t)$ be orthogonal on $[0, 1]$ with respect to the weight function $p(t) > 0$. Make the Liouville transformation and show that $f(t)$, $g(t)$ are orthogonal on an appropriate interval $[0, c]$.

2. Carry out the change of boundary condition from (6.3) into (6.12).

3. Give a direct proof of Theorem 5.1 for the simpler system (6.10)–(6.12).

4. Consider the boundary value problem

$$y'' + (\lambda - q(t))y = 0$$

$$y(0) = y(1)$$

$$y'(0) = y'(1).$$

If $y_1(t)$, $y_2(t)$ are solutions of this problem on $(0, 1)$ for distinct values $\lambda = \lambda_1$ and $\lambda = \lambda_2$, show that these solutions are orthogonal.

5. Consider the boundary value problem

$$y^{(iv)} + y'' + (\lambda - q(t))y = 0$$

$$y(0) = y'(0) = 0$$

$$y(1) = y'(1) = 0.$$

If $y(t, \lambda_1)$ and $y(t, \lambda_2)$ are solutions of this problem on $(0, 1)$ for distinct values of $\lambda = \lambda_1$ and $\lambda = \lambda_2$, show that these solutions are orthogonal.

7. An Alternate Formulation—Integral Equations

Consider the Sturm-Liouville problem

$$y'' + [\lambda - q(t)]y = 0 \tag{7.1}$$

$$y(a) = 0 \tag{7.2}$$

$$y(b) = 0. \tag{7.3}$$

Equation (7.1) can be rewritten as

$$y'' - q(t)y = -\lambda y \tag{7.4}$$

and the right-hand side viewed as a forcing term for the differential equation

$$y'' - q(t)y = 0. \tag{7.5}$$

Let $G(t, \tau)$ be the Green's function introduced in Section 2; that is,

$$G(t, \tau) = \begin{cases} \dfrac{\varphi(t)\psi(\tau)}{W(\varphi, \psi)(\tau)} & a \le t \le \tau \\[2ex] \dfrac{\varphi(\tau)\psi(t)}{W(\varphi, \psi)(\tau)} & \tau \le t \le b, \end{cases}$$

where $\varphi(t)$ and $\psi(t)$ are linearly independent solutions of (7.5) with $\varphi(a) = 0$, $\psi(b) = 0$, and W is the Wronskian of $\varphi(t)$ and $\psi(t)$. In fact, for (7.5) the Wronskian will be constant, since the first derivative term does not appear. Following the proof of Theorem 2.2 it is easy to show that a solution of (7.1)–(7.3) is given by

$$y(t) = \lambda \int_a^b G(t, \tau)y(\tau)\,d\tau \tag{7.6}$$

(where, of course, $y(t)$ is unknown). Equation (7.6) is an *integral equation* for an unknown function $y(t)$. This is an object of considerable mathematical study and perhaps the most elegant way to present the Sturm-Liouville theory. If we write $\mu = 1/\lambda$ (assume that $\lambda > 0$) and define a linear operator T by

$$(T\varphi)(t) = \int_a^b G(t, \tau)\varphi(\tau)\,d\tau,$$

then an equivalent problem is to find values of μ so that

$$Ty = \mu y \tag{7.7}$$

has a nontrivial solution. Operator equations such as (7.7) also have a very complete theory that includes, as a special case, the Sturm-Liouville theory. The necessary mathematical concepts required to make use of this are well beyond the prerequisites for this course, so they will not be pursued here. These ideas are introduced because the student may find this formulation in other books.

8. Eigenfunction Expansions

In Section 6 it was shown that eigenfunctions corresponding to distinct eigenvalues of a Sturm-Liouville problem are orthogonal. In this section we show

how to exploit this property of orthogonality to express an arbitrary function as the sum of a series of eigenfunctions. This technique will be applied to inhomogeneous eigenvalue problems and to some problems in partial differential equations. These ideas and techniques form one of the cornerstones of analysis, but it will be possible to see only the "tip" of this very beautiful and very deep theory. To make the mathematics entirely rigorous requires more analysis than has been assumed in this text. Traditionally, the technique has been taught with a combination of analogy and intuitive ideas, and this approach will be pursued here.

We will move away from our primary subject to explore some basic ideas in analysis, and will return in the following sections to apply these ideas to differential equations. The concept of a linear space, and a basis for it, occurred in the first chapter. Our early results on linear equations can be summarized in the statement that solutions of

$$x' = Ax$$

($x \in R^n$, A an $n \times n$ matrix) span an n-dimensional vector space, and, thus, finding n linearly independent solutions allows us to find all solutions (as linear combinations of the linearly independent ones). These solutions are part of a larger space, the space of continuously differentiable functions defined on R^n. It is not possible to describe all of the continuously differentiable functions by a linear combination of any finite collection of functions. Might it be possible to describe all of the functions as an "infinite linear combination" of functions from a "smaller," select set?

To establish the idea, consider the set of all continuous functions on the interval [0, 1] and use the metric introduced in Chapter 3, namely,

$$d(x, y) = \max_{[0, 1]} |x(t) - y(t)|.$$

We have previously denoted this metric space by $C[0, 1]$. This space contains all of the polynomials and, in particular, the set of functions $\{1, x, x^2, x^3, \ldots\}$, which we shall call $P[0, 1]$. A set like $\{1, x, x^2, x^3, \ldots\}$, which can be put into one-to-one correspondence with the positive integers (one function for each integer and one integer for each function in the set) is said to be *countable*. The set $C[0, 1]$ contains functions that cannot be written as a finite linear combination of elements of P, that is, cannot be written as

$$\sum_{i=0}^{n} c_i x^i,$$

for some value of n. An example of such a function would be any function that does not have a first derivative, or a function like $\sin(x)$. However, the following result is true.

THEOREM 8.1 *(Weierstrass approximation theorem)*
    Let f be an element of $C[0, 1]$. Given any $\varepsilon > 0$, there is an element p, which is a finite linear combination of elements in $P[0, 1]$, such that $d(f, p) < \varepsilon$, where d is the metric in $C[0, 1]$.

This theorem may be summarized by saying that every continuous function on $[0, 1]$ can be "approximated" by a polynomial. If every element of a metric space has an element of a particular subset arbitrarily close (in the metric of the space) to it, the subset is said to be *dense* in the space. The preceding theorem says that the polynomials are dense in $C[0, 1]$.

It is often convenient to approximate by functions other than polynomials, and it is often necessary to deal with functions that are not continuous. (In particular, piecewise continuous functions are needed in engineering problems.) We will develop a space larger than the space $C[0, 1]$ and countable sets of functions such that linear combinations of elements of the set are dense in the space. These functions will play the role that basis vectors play in Euclidean space—every function in the space will be represented as a linear combination (in a limiting sense) of elements of this countable set. Specifically, the end result will be to write an arbitrary function in the space as a sum,

$$f(x) = \sum_{i=0}^{\infty} c_i \varphi_i,$$

where equality will mean the limit of the sequence of partial sums in an appropriate metric. To accommodate a wider class of functions than continuous ones, we first develop a suitable "space" in which our functions are to reside.

Recall that a metric space consists of two items, a set and a metric. In Chapter 3 we noted that the same set with two different metrics or different sets with the same metric constitute different spaces. For example, the points in the plane (x, y) can be given different metrics. Let $P_i = (x_i, y_i)$. We can assign the metric

$$d(P_1, P_2) = [(x_1 - x_2)^2 + (y_1 - y_2)^2]^{1/2}$$

or the metric

$$d(P_1, P_2) = |x_1 - x_2| + |y_1 - y_2|.$$

The first is the familiar Euclidean metric and the second is the "taxicab" metric, which was used in Chapter 3. As an example of different sets with the same metric, we noted the real numbers and the rational numbers, each with the metric of absolute value of the difference of the numbers. It will be helpful to the reader to keep these simple examples in mind in what follows.

We take as the basic set the set of all continuous functions on an interval [0, 1]. We assign to pairs of such functions the distance, or metric, given by

$$d(x, y) = \int_0^1 |x(t) - y(t)| \, dt. \tag{8.1}$$

This set and this metric form a metric space. Since the functions are continuous and the integrand is positive, the integral in (8.1) is nonnegative and is zero only if the two functions are identical for every point in the interval. Moreover, interchanging x and y yields the same number. The triangle inequality follows from the same property for absolute value and the additivity of the integral. The result of assigning this metric to the continuous functions on a closed interval produces a different metric space from $C[0, 1]$ used in Chapter 3. The metric space has the disadvantage that it is not complete. Recall that this means that there are convergent sequences of elements of the space that do not have a limit in the space. It will be a completion process that will widen the class of functions beyond continuous ones in our study. First, however, we consider the general completion problem.

Consider two metric spaces M_1 and M_2 with metrics d_1 and d_2. We say that the metric space M_1 is *embedded* in M_2 if (as a set of points) M_1 is a subset of M_2 and if for each pair of points x and y in M_1 it is the case that $d_1(x, y) = d_2(x, y)$. The basic idea is that M_2 is a larger space but that the definitions of distance agree for pairs of points that are in the smaller set. The easiest case is when the metrics are the same; for example, the rational numbers are a subset of the real numbers and the same distance function, $|x - y|$, is used for both. The metric space we have is not complete, so a natural question is whether there is a larger metric space containing it that is complete and in which the given one is embedded. The answer to this question is always yes, as the following theorem states.

THEOREM 8.2

Every metric space may be embedded, as a dense subset, in a complete metric space.

The proof of the theorem is beyond the scope of this course, but it may be found in most standard topology books (see Dugundji (1970), page 304, for example). The process is analogous to obtaining the real numbers by completing the rational numbers. It may be viewed as adding something to fill up the "holes," just as irrational numbers fill up the "holes" in the set of rational numbers. The new space obtained by completing the continuous functions with the metric (8.1) will be denoted by $L^1[0, 1]$. This is the space of "Lebesque integrable functions" on $[0, 1]$. We will not need this space, nor will we discuss Lebesque integration, but it is a way station on the path to the space we do want. It will serve to illustrate some peculiarities of the process and is an important space for other purposes.

There is a certain mystery about the new space created by the extension process, because it is not entirely clear what has been added. The elements added to the space—sometimes called "ideal" elements—are not known explicitly, and need not be known for our purposes. It is similar to computer arithmetic, where only rational numbers exist but π can always be approximated. In the same way the ideal elements of $L^1[0, 1]$ can always be approximated in the metric by continuous functions. Any properties to be established about these added functions have to be established by taking limits of continuous functions. We will carefully skirt this issue whenever possible by dealing with continuous or obvious limits of continuous functions. When taking a limit, we must be careful not to assert that the limiting element is a continuous function. We note the following.

1. All Riemann integrable functions (the functions found in calculus courses) are in $L^1[0, 1]$. In particular, all piecewise continuous functions defined on $[0, 1]$ are in $L^1[0, 1]$.

2. Any linear combination of functions in $L^1[0, 1]$ is in $L^1[0, 1]$.

Now take the same continuous functions on $[0, 1]$ and introduce the metric

$$d(x, y) = \left[\int_0^1 |x(t) - y(t)|^2 \, dt \right]^{1/2}. \tag{8.2}$$

This function is nonnegative, equal to zero if and only if the functions $x(t)$ and $y(t)$ are the same and $d(x, y) = d(y, x)$. To show that it forms a metric space requires only to establish the triangle inequality. We omit this (computational) point. Again, this metric space is not complete and it may be embedded in a complete metric space using Theorem 8.2. This metric space is called $L^2[0, 1]$ and is the space that is needed for our theory. This space contains elements that are not continuous functions, but we can avoid having to deal with them by working with limits of continuous functions. The above two properties remain valid, although the two sets of functions are not identical.

To avoid ambiguities later, we pause here to note the different ways in which the idea of convergence enters. Let $x_n(t)$ be a sequence of continuous functions defined on the interval $[0, 1]$. What do we mean by

$$\lim_{n \to \infty} x_n = y?$$

If we view the sequence $x_n(t)$ as just a sequence of functions, we mean that for each fixed t in the interval, $\lim_{n\to\infty} x(t)$ exists and the limit is given by the value $y(t)$. This is called *pointwise convergence*. If we view the sequence x_n as a sequence of elements in $C[0, 1]$, we mean that the functions converge in the metric of that space to a continuous function y, also in the space. This is called *uniform convergence*. Uniform convergence implies pointwise convergence. If we view the sequence x_n as a sequence of elements of $L^2[0, 1]$, then the sequence converges in the metric of that space to an element of that space. This is called *mean square convergence*. Note that, in this case, the limit might be one of the ideal elements of that space and not a continuous function. Uniform convergence implies mean square convergence.

To illustrate the difference, let us consider a simple example. Let

$$f_n(t) = t^n \quad \text{for } 0 \le t \le 1.$$

The sequence converges pointwise to

$$f(t) = \begin{cases} 0, & t \ne 1 \\ 1, & t = 1, \end{cases}$$

since the sequence of real numbers $f_n(t)$ converges to the indicated values for each t. If the functions are viewed as points in $C[0, 1]$, the sequence does not converge in the metric of that space. (In particular, it cannot converge to $f(t)$ as just defined, since that function is not continuous.) Now let $h(t) = 0$ for each t in $[0, 1]$. In the space $L^2[0, 1]$, $\lim_{n\to\infty} f_n = h$, since

$$d(f_n, h) = \left[\int_0^1 t^{2n}\, dt \right]^{1/2} = \left(\frac{1}{2n + 1} \right)^{1/2} \to 0 \quad \text{as} \quad n \to \infty.$$

Thus the three concepts are different.

Next we define a number called an *inner product*, by

$$\langle x, y \rangle = \int_0^1 x(t) y(t)\, dt.$$

We shall make use of this concept mostly for notational purposes, but it is an important concept in mathematics in other connections. When $x(t)$ and $y(t)$ are continuous functions and α is a real number, some obvious properties of the inner product are

1. $\langle x, x \rangle \ge 0$ and $\langle x, x \rangle = 0$ if and only if $x = 0$ for all t

2. $\langle x, y \rangle = \langle y, x \rangle$

3. $\langle x + y, z \rangle = \langle x, z \rangle + \langle y, z \rangle$

4. $\langle \alpha x, y \rangle = \alpha \langle x, y \rangle$.

A less-than-obvious consequence is the following.

THEOREM   (*Schwarz inequality*)

$$|\langle x, y \rangle| \leq \langle x, x \rangle^{1/2} \langle y, y \rangle^{1/2}.$$

Proof: The expansion of the nonnegative quantity $\langle x + \alpha y, x + \alpha y \rangle$ yields

$$\alpha^2 \langle y, y \rangle + 2\alpha \langle x, y \rangle + \langle x, x \rangle \geq 0.$$

This is a quadratic in the variable α, so the discriminant must be non-positive for it to remain nonnegative. We have that

$$\langle x, y \rangle^2 \leq \langle x, x \rangle \langle y, y \rangle,$$

which leads to the desired inequality.

Note that $\langle x, x \rangle$ is the square of the distance of the element x from the "origin"—from the identically zero function. With this interpretation, one immediate consequence of the Schwarz inequality is that the product of two functions that are in $L^2[0, 1]$ is in $L^1[0, 1]$. Note also that the condition of orthogonality (with weight function identically 1) of two functions $x(t)$ and $y(t)$ can be simply expressed as

$$\langle x, y \rangle = 0.$$

Now let φ_n, $i = 0, 1, \ldots$ be a sequence of nontrivial, continuous functions such that $\langle \varphi_i, \varphi_j \rangle = 0$ if $i \neq j$. This is called an orthogonal sequence. Without loss of generality, we can also assume that $\langle \varphi_i, \varphi_i \rangle = 1$. This can be accomplished by multiplying each function by an appropriate constant. This is just a normalization; that is, it makes the norm in L^2 equal to 1. The resulting sequence is said to be *orthonormal*. View this sequence of functions as "coordinates" in the same way that we view unit coordinate vectors in R^n. Choose a function from $L^2[0, 1]$—a continuous (or piecewise continuous) one. We now ask, how can we choose a linear combination of the first n elements of the orthonormal set so as to get as close as possible (in the $L^2[0, 1]$ metric) to the function f? Phrased another way, choose constants c_i so that

$$d\left(f, \sum_{i=0}^{n-1} c_i \varphi_i\right)$$

is minimized. Restated once more, find the point in the span of the first n elements of the orthonormal set that is closest to f in $L^2[0, 1]$. Since the set is finite, this is a simple calculus problem. We obtain a necessary condition by differentiating

$$\int_0^1 \left[f - \sum_{i=0}^{n-1} c_i \varphi_i \right]^2 dt$$

with respect to c_i and setting the result equal to zero. Expanding within the square brackets and integrating yields (using inner-product notation)

$$\langle f, f \rangle - 2 \sum_{i=0}^{n-1} c_i \langle \varphi_i, f \rangle + \sum_{i=0}^{n-1} c_i^2.$$

Differentiation yields

$$-2\langle f, \varphi_i \rangle + 2c_i = 0, \qquad i = 1, \ldots, n,$$

so that a necessary condition is that

$$c_i = \langle f, \varphi_i \rangle.$$

It is not difficult to argue that the minimum is attained for this choice.

We might suspect that as more functions from the orthonormal set are allowed, the approximation should improve. The principal question is whether or not

$$\lim_{n \to \infty} d\left(f, \sum_{i=0}^{n} c_i \varphi_i \right) = 0. \tag{8.3}$$

If so, then f has been approximated in the space $L^2[0, 1]$. When this happens, it is customary to write

$$f = \sum_{i=0}^{\infty} c_i \varphi_i. \tag{8.4}$$

Whether the limit (8.3) is zero really depends on the extent of the span of the orthonormal set. If coefficients c_i can be found so that (8.4) holds for every function in $L^2[0, 1]$ (and hence for all elements there), then the *orthonormal set* is said to be *complete*. The principal result we wish to state concerns the completeness of the eigenfunctions corresponding to the Sturm-Liouville problem studied in Section 6.

THEOREM 8.3

Eigenfunctions corresponding to the Sturm-Liouville problem

$$w'' - (\lambda - Q(x))w = 0 \tag{8.5}$$

$$\gamma_1 w(0) + \gamma_2 w'(0) = 0 \tag{8.6}$$

$$\delta_1 w(1) + \delta_2 w'(1) = 0 \tag{8.7}$$

can be chosen to form a complete orthonormal set in $L^2[0, 1]$.

THEOREM 8.4

Let φ_n be the complete orthonormal set given by the eigenfunctions of
(8.5)–(8.7), and let f be an element of $L^2[0, 1]$. Then

$$f = \sum_{i=0}^{\infty} c_i \varphi_i,$$

where $c_i = \langle f, \varphi_i \rangle$ and where convergence of the series is interpreted as mean
square convergence of the partial sums.

There are many applications of the idea of expansion of a function into components of a complete orthonormal set. The correct perspective is to view the functions φ_n as coordinates or "basis elements" in an "infinite dimensional space."
We illustrate the theorem with some simple examples for a special eigenvalue
problem that lies at the beginning of the theory of Fourier series.

We begin with the problem considered in Section 4,

$$y'' + \lambda y = 0$$

$$y(0) = 0, \qquad y(1) = 0.$$

The eigenvalues were $\lambda = n^2 \pi^2$, $n = 1, 2, \ldots$, and the eigenfunctions were of
the form $\sin(n\pi x)$. To effect the normalization, we take the orthonormal sequence to be $\varphi_n = \sqrt{2} \sin(n\pi x)$. First, consider the function $f(x) = x, 0 \le x \le 1$.
Then we have

$$x = \sum_{n=1}^{\infty} c_n \sqrt{2} \sin(n\pi x)$$

where

$$c_n = \langle x, \sqrt{2} \sin(n\pi x) \rangle = \int_0^1 \sqrt{2} x \sin(n\pi x)\, dx$$

$$= (-1)^{n+1} \frac{\sqrt{2}}{n\pi}.$$

Thus (8.4) takes the form

$$x = \frac{2}{\pi}\left[\sin(\pi x) - \frac{1}{2}\sin(2\pi x) + \cdots \right].$$

The series converges in the mean square sense; that is, it converges in the metric of $L^2[0, 1]$.

For a more interesting case, take $f(x) = e^x$. Then

$$\langle e^x, \sqrt{2}\sin(n\pi x)\rangle = \sqrt{2}\int_0^1 e^x \sin(n\pi x)\, dx$$

$$= \frac{\sqrt{2}n\pi[1 - e\cos(n\pi)]}{1 + \pi^2 n^2}.$$

Thus the expansion takes the form

$$e^x = (2\pi)\left[\frac{(1 + e)\sin(x)}{1 + \pi^2} + \frac{2(1 - e)\sin(2x)}{1 + 4\pi^2} + \cdots \right].$$

These expansions are known as the *Fourier sine series* of the given function.

A similar eigenvalue problem is

$$y'' + \lambda y = 0$$
$$y'(0) = 0, \qquad y'(1) = 0.$$

The general solution to the differential equation is $y = c_1 \sin(\sqrt{\lambda}x) + c_2 \cos(\sqrt{\lambda}x)$. To satisfy the first boundary condition requires that $c_1 = 0$, and to satisfy the second requires that

$$\sqrt{\lambda}c_2 \sin(\sqrt{\lambda}) = 0$$

or that

$$\lambda = n^2\pi^2.$$

The orthonormal system may be taken to be

$$\varphi_0 = 1$$
$$\varphi_n = \sqrt{2}\cos(n\pi x), \qquad n = 1, 2, 3, \dots.$$

It is important to note that there now is a nontrivial term corresponding to $n = 0$, a constant term. This will require a special computation to be made for the case $n = 0$. Let us again expand the preceding two functions, but this time in terms of the eigenfunctions of the last Sturm-Liouville problem. A similar computation yields

$$c_0 = \langle x, 1 \rangle = \frac{1}{2}$$

$$c_n = \langle x, \sqrt{2} \cos(n\pi x) \rangle$$

$$= \frac{\sqrt{2}[(-1)^n - 1]}{n^2 \pi^2}, \qquad n = 1, 2, 3, \ldots.$$

Hence, the expansion takes the form

$$x = \frac{1}{2} - \frac{4}{\pi^2}\left(\cos(\pi x) + \frac{1}{9}\cos(3\pi x) + \cdots\right).$$

This is called the *Fourier cosine expansion*.

In a similar way we can compute the Fourier cosine expansion of e^x and find

$$e^x = e - 1 - 2\left[\frac{1+e}{1+\pi^2}\cos(\pi x) + \frac{1-e}{1+4\pi^2}\cos(2\pi x) + \cdots\right].$$

Odd functions are those for which $f(-x) = -f(x)$ and even functions are those for which $f(-x) = f(x)$. The sine series consists only of odd functions (sines); hence, if the expansion above is viewed on the larger interval $[-1, 1]$, using the same coefficients, it must represent an odd function on this interval. Hence on $[-1, 1]$ the Fourier sine series represents x. However, the cosine series consists of even functions and if it is considered on $[-1, 1]$ it will represent an even function, $|x|$. Figure 8.1 illustrates this point.

Every function can be written as the sum of an even and an odd function. Specifically, if $f(x)$ is the given function, then $f(x) = f^o(x) + f^e(x)$, where

$$f^o(x) = \frac{f(x) - f(-x)}{2}$$

$$f^e(x) = \frac{f(x) + f(-x)}{2}.$$

We obtain the *Fourier series* of $f(x)$ on $[-1, 1]$ by breaking it into its odd and even parts and finding the sine and cosine series, respectively. For example, we can write

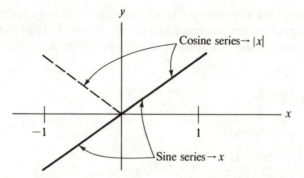

Figure 8.1 Convergence on $[-1, 1]$ of the Fourier sine and cosine series for the function $f(x) = x$ on $[0, 1]$. The sine series converges to the odd extension of x and the cosine series converges to the even extension of x.

$$e^x = \frac{1}{2}(e^x + e^{-x}) + \frac{1}{2}(e^x - e^{-x})$$

$$= \cosh(x) + \sinh(x).$$

Thus we could find the Fourier sine series for $\sinh(x)$ and the Fourier cosine series for $\cosh(x)$ and combine them to find the Fourier series for e^x. Of course it is not difficult to show that the set $[1, \sin(n\pi x), \cos(n\pi x), i = 1, 2, \ldots]$ is orthogonal on $[-1, 1]$ and hence obtain the expansion directly in $L^2[-1, 1]$. Completeness would not follow from Theorem 8.3, since this orthonormal set was not obtained from a Sturm-Liouville problem of the type we have been considering. It is complete, however.

In applications we wish to manipulate the series given by Theorem 8.4, in particular to do term-by-term operations, as we can with a Taylor series. We cannot give an adequate explanation of the exact terms under which such operations are valid without further analysis. Mean square convergence does justify integration on a term-by-term basis and we note (without proof) the following theorem, which is the beginning of the "smooth" theory (as opposed to the L^2 theory that we just indicated).

THEOREM 8.5

If f has a continuous second derivative on $[0, 1]$ and satisfies the boundary conditions (8.6) and (8.7), then the series given in Theorem 8.4 converges in the metric of $C[0, 1]$.

EXERCISES

1. Find the Fourier sine series expansion of $\sinh(x)$ and $\cosh(x)$ on $[0, 1]$.

2. Find the Fourier cosine series expansion of $\sinh(x)$ and $\cosh(x)$ on $[0, 1]$.

3. Find the Fourier sine series expansion of $\cos(x)$ on $[0, 1]$.

4. Find the Fourier series expansion of e^x on $[-1, 1]$.

5. Give an example of a piecewise continuous function that is in $L^1[0, 1]$ but is not in $L^2[0, 1]$.

6. Determine whether the following functions are odd, even, or neither: $x - x^2$, $x \sin(x)$, $x \cos(x)$, $x^2 \cos(x)$, e^{x^2}, $|x|^3$.

7. Define a sequence of piecewise continuous functions $f_n(x)$ by

$$f_n(x) = \begin{cases} 0, & x = 0 \\ n, & 0 < x < \dfrac{1}{n} \\ 0, & \dfrac{1}{n} \le x \le 1. \end{cases}$$

Investigate pointwise and mean square convergence to the zero function.

8. Repeat Exercise 7 for

$$f_n(x) = \begin{cases} 0, & x = 0 \\ n^{1/2}, & 0 < x < \dfrac{1}{n} \\ 0, & \dfrac{1}{n} \le x \le 1. \end{cases}$$

9. Repeat Exercise 7 for

$$f_n(x) = \begin{cases} 0, & x = 0 \\ 1, & 0 < x < \dfrac{1}{n} \\ 0, & \dfrac{1}{n} \le x \le 1. \end{cases}$$

10. Repeat Exercises 7, 8, and 9, with mean square convergence replaced by convergence in the metric of $L^1[0, 1]$.

11. If

$$f(x) = \sum_1^\infty c_n \sin(n\pi x),$$

show formally that

$$\langle f(x), f(x) \rangle = \sum_{1}^{\infty} c_n^2.$$

12. Find the Fourier sine series for

$$f(x) = \begin{cases} x, & 0 \le x \le \dfrac{1}{2} \\ 1 - x, & \dfrac{1}{2} \le x \le 1. \end{cases}$$

13. Find the Fourier cosine series for the function in Exercise 12 and compare the extensions of each series to $[-1, 1]$.

14. Find the Fourier sine and cosine series for

$$f(x) = \begin{cases} 1, & 0 \le x \le \dfrac{1}{2} \\ 0, & \dfrac{1}{2} < x \le 1. \end{cases}$$

15. Find the Fourier series on $[-1, 1]$ of

$$f(x) = \begin{cases} 0, & -1 \le x \le 0 \\ 1, & 0 < x \le 1. \end{cases}$$

9. The Inhomogeneous Sturm-Liouville Problem

The eigenfunction expansions developed in the previous section have a great many applications. We illustrate one of them here by considering the inhomogeneous (or forced) Sturm-Liouville problem

$$y'' + (\lambda - Q(t))y = f(t)$$
$$y(0) = 0 \tag{9.1}$$
$$y(1) = 0,$$

where f and Q are continuous functions (weaker hypotheses are possible and stronger ones are convenient). We proceed in a purely formal manner, omitting the mathematical details. The basic idea is to "find" a representation of the solution of the problem in a series of orthogonal functions; that is, the solution is "found" by finding its eigenfunction expansion. As usual with forced problems, we assume that everything is known about the solution of the unforced or homogeneous case,

$$y'' + (\lambda - Q(t))y = 0$$
$$y(0) = 0 \tag{9.2}$$
$$y(1) = 0.$$

In particular, we know that, for this problem, there exists an infinite sequence of eigenvalues, λ_n, and an infinite sequence of eigenfunctions, φ_n, $n = 1, 2, \ldots$. If there exists a solution of the problem (9.1), then it is twice-differentiable and hence can be expanded in the eigenfunctions of the problem (9.2). (We tacitly pass over the question of the existence of a solution for the moment.) Thus, if y denotes the solution of (9.1), then we write

$$y = \sum_1^\infty c_n \varphi_n. \tag{9.3}$$

The function $f(t)$ can also be expanded in a series of eigenfunctions to obtain

$$f = \sum_1^\infty C_n \varphi_n. \tag{9.4}$$

We now proceed to a little trickery with the differential equations in (9.1) and (9.2) and the series (9.3) and (9.4). Substitute the series (9.3) into the equation (9.1). Assume that the second derivative of the series may be obtained by differentiating each term of the series twice. This would be valid if Q and f were sufficiently differentiable. Then

$$\left(\sum_1^\infty c_n \varphi_n \right)'' = \sum_1^\infty c_n \varphi_n''$$

yields that

$$\sum_1^\infty [c_n \varphi_n'' + (\lambda - Q)c_n \varphi_n] = \sum_1^\infty C_n \varphi_n. \tag{9.5}$$

However,

$$\varphi_n'' - Q\varphi_n = -\lambda_n \varphi_n, \tag{9.6}$$

since φ_n is a solution of (9.2). Hence, replacing terms in (9.5) by (9.6) produces

$$\sum_1^\infty c_n(\lambda - \lambda_n)\varphi_n = \sum_1^\infty C_n \varphi_n. \tag{9.7}$$

The coefficients, C_n, are known, since f is a given function—"known" in the sense that they can be computed by the methods of the previous section. The c_n are the unknown coefficients in the expansion of the solution in terms of the orthogonal eigenfunctions of the homogeneous problem. Equating coefficients in (9.7) yields

$$c_n = \frac{C_n}{\lambda - \lambda_n}.$$

Hence we have the coefficients in the expansion of the solution of (9.1) in terms of the eigenfunctions of the homogeneous problem (9.2).

$$y(t) = \frac{C_0}{\lambda - \lambda_0} \varphi_0 + \frac{C_1}{\lambda - \lambda_1} \varphi_1 + \frac{C_2}{\lambda - \lambda_2} \varphi_2 + \cdots. \tag{9.8}$$

Care must be taken in the interpretation of the formula above, since convergence and smoothness questions (the resulting $y(t)$ must be twice-differentiable, for example) remain unanswered. Treating this formally obtained result as accurate suggests a very different behavior for Sturm-Liouville problems with forcing than without. The formula is meaningful if $\lambda \neq \lambda_n$, that is, if λ is not equal to an eigenvalue of the Sturm-Liouville problem (9.2). As $\lambda \to \lambda_n$, solution (9.8) will not exist unless $C_n = 0$. This will happen only if $\langle \varphi_n, f \rangle = 0$, or, if φ_n is orthogonal to f. This result is, in fact, a theorem.

THEOREM 9.1

Fix $\lambda \neq 0$. If λ is not an eigenvalue of the Sturm-Liouville problem (9.2), then (9.1) has a unique solution whose eigenfunction expansion is given by (9.8). If λ is an eigenvalue of (9.2), then (9.1) can have a solution only if $\langle y, f \rangle = 0$ for every solution $y(t)$ of (9.2).

It is not possible for us to make this statement entirely rigorous, but the first part does follow from Theorem 2.2, developed earlier. We restate a portion of this in the form of an alternative theorem.

THEOREM 9.2

If, for a fixed value of λ, (9.2) has only the trivial solution, then (9.1) has a unique solution.

We illustrate the technique on a simple problem,

$$y'' + \lambda y = e^t$$
$$y(0) = 0 \qquad\qquad\qquad\qquad\qquad\qquad\qquad (9.9)$$
$$y(1) = 0.$$

The unforced problem has normalized eigenfunctions $\varphi_n = \sqrt{2}\sin(n\pi t)$ corresponding to $\lambda_n = n^2\pi^2$. Hence if $\lambda \neq \lambda_n$, then there will be a unique solution of the boundary value problem (9.9). To compute the eigenfunction expansion of the solution, recall that

$$e^t = \sum_1^\infty \frac{(2n\pi)(1 - e(-1)^n)}{1 + n^2\pi^2}\sin(n\pi t).$$

Hence $y(t, \lambda)$, the solution of (9.9) for this fixed value of λ, has the expansion

$$y(t, \lambda) = \sum_1^\infty \frac{(2n\pi)(1 - e(-1)^n)}{(1 + n^2\pi^2)(\lambda - n^2\pi^2)}\sin(n\pi t).$$

Since this problem has constant coefficients in the homogeneous part and a simple function on the right-hand side, it is possible to find the solution directly. The power of this approach is seen in solving more complicated problems with variable coefficients (with a $Q(t)$ term). Many of the special functions of mathematical physics arise as solutions of Sturm-Liouville problems, generating complete orthogonal sets. The user must choose the proper orthonormal sequence for his or her particular problem. The theory itself is a powerful analytic tool.

EXERCISES

1. Find the eigenfunction expansion of the forced boundary value problem (9.9), where e^t is replaced by

   (a) e^{-t}

   (b) $\sinh(t)$

   (c) $\cosh(t)$

   (d) $e^{\pi t}$

2. Carry out the development for solutions of (9.1) for the Sturm-Liouville problem

$$y'' + (\lambda - Q(t))y = f(t)$$

$$y(0) = 0, \qquad y'(0) = 0.$$

3. Find the eigenfunction expansion of the forced boundary value problem

$$y'' + \lambda y = f(t)$$

$$y(0) = 0, \qquad y'(0) = 0$$

where $f(t)$ is

(a) e^t

(b) $\cosh(t)$

4. Express the solution of (9.9) in terms of the Green's function found in Section 2. Carry out the integration to find an exact solution.

10. Some Standard Applications of Sturm-Liouville Theory

When Sturm-Liouville problems were introduced, it was noted that these problems were useful in the solution of partial differential equations. One technique, in particular, for solving linear partial differential equations leads to Sturm-Liouville problems in a very natural way. We illustrate this approach here with two of the simplest such problems. However, since this course does not include a study of partial differential equations, we shall merely exhibit the appropriate model without a detailed explanation, show the solution technique, and pass on without much further comment. Many mathematical details are necessarily omitted. The discussion is intended only to illustrate the way in which the eigenvalue problem and the expansion in orthogonal eigenfunctions play a prominent role.

The first application considered is commonly called the "plucked string." Our model is an elastic "string," of uniform composition, stretched between two points, which we take to be one unit apart on the x-axis (see Figure 10.1). If the string is deformed at time $t = 0$ and released, then we anticipate that the string will "vibrate." We assume that these vibrations can be described by showing how the size of the displacement from the axis, at the point x, varies as a function of the time t. Denote this quantity by $u(x, t)$. For the plucked string we have in mind an initial deviation of the form shown in Figure 10.2 (an analytic description will follow). Standard physical arguments lead us to expect that this deviation can be approximated by a solution of the partial differential equation

$$u_{xx}(x, t) = \left(\frac{1}{a}\right)^2 u_{tt}(x, t), \tag{10.1}$$

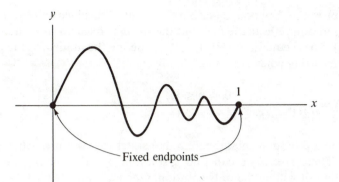

Figure 10.1 A deformed elastic string with ends held fixed at $x = 0$ and $x = 1$.

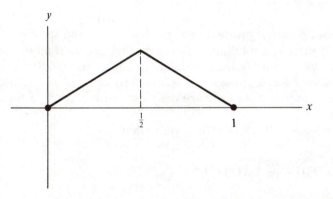

Figure 10.2 The initial deformation corresponding to a plucked string.

called the *wave equation*. The constant a is a function of the type of material and the construction of the string. The fact that the string is fixed at the endpoints for all times (for all values of t) adds the conditions

$$u(0, t) = 0, \qquad t > 0, \tag{10.2}$$

$$u(1, t) = 0, \qquad t > 0. \tag{10.3}$$

These are called *boundary conditions*. Release from the initial displacement at time zero provides initial conditions. These are expressed by

$$u(x, 0) = f(x), \tag{10.4}$$

$$u_t(x, 0) = 0, \qquad 0 \leq x \leq 1, \tag{10.5}$$

where $f(x)$ is the function describing the initial displacement and where the second condition reflects the fact that the string is released (no initial velocity is imparted). These equations [(10.1)–(10.5)] define the model for the problem of the plucked string when $f(x)$ is defined by

$$f(x) = \begin{cases} 2x, & 0 \le x \le \frac{1}{2}, \\ 2(1-x), & \frac{1}{2} < x \le 1. \end{cases} \tag{10.6}$$

There is a fundamental theorem that states that this problem has a unique solution. Therefore, if we can find a solution by any technique—under any assumption—it will be the only solution. One technique, called "separation of variables," is to assume that a solution has the form

$$u(x, t) = X(x)T(t). \tag{10.7}$$

The assumption is not justified here, but we can try and see if it does lead to a solution. If there is a solution of this form that can meet all of the prescribed conditions, then it must be the only solution. The form of (10.7) makes the partial derivatives very simple, for $u_x(x, t) = X'(x)T(t)$ and $u_t(x, t) = X(x)T'(t)$, where the prime, $'$, indicates an ordinary derivative with respect to the appropriate argument.

Substituting (10.7) into (10.1) yields that

$$X''(x)T(t) = \left(\frac{1}{a}\right)^2 X(x)T''(t)$$

or, being careless about where these functions might be zero, that

$$\frac{X''(x)}{X(x)} = \left(\frac{1}{a}\right)^2 \frac{T''(t)}{T(t)}.$$

We argue that the left-hand side is independent of the variable t and that the right-hand side is independent of the variable x, but since they are equal for all t and x, and t and x are independent, the common value must be constant. Call the constant $-k$ (the minus sign anticipates how things will work out, eventually). This means that X and T each satisfy ordinary differential equations of the form

$$X'' + kX = 0$$
$$T'' + a^2 kT = 0.$$

The constraints at the endpoints (10.2) and (10.3) lead to the conclusion that $X(x)$ must be a solution of the Sturm-Liouville problem

$$X'' + kX = 0$$
$$X(0) = 0 \qquad\qquad (10.8)$$
$$X(1) = 0$$

where k is a parameter. Thus, from our previous study, we know that there will be a nontrivial solution only if

$$k = n^2\pi^2.$$

Thus,

$$X(x) = C\sin(n\pi x),$$

where C is any constant.

Using this value of k in the equation for T and the condition (10.5) says that T is a solution of

$$T'' + (an\pi)^2 T = 0$$
$$T'(0) = 0. \qquad\qquad (10.9)$$

A simple computation yields that $T(t)$ must be of the form

$$T(t) = B\cos(an\pi t),$$

where B is an arbitrary constant.

Putting these two specifications into (10.7) tells us that a solution of (10.1), (10.2), (10.3), (10.5) takes the form

$$u(x, t) = A_n \sin(n\pi x)\cos(an\pi t). \qquad\qquad (10.10)$$

The arbitrary constants C and B have been combined into the constant A_n, the n denoting that we can choose a different A for each choice of n; that is, (10.10) is a solution for every integer n and for every choice of the constant A. Further, any finite linear combination of solutions of the form (10.10) will again be a solution. Thus

$$\sum_1^N A_n \sin(n\pi x)\cos(an\pi t) \qquad\qquad (10.11)$$

is a solution for every choice of constants A_n, $n = 1, 2, \ldots, N$. Might it be the case that we could let the number of terms in the linear combination go to infinity and find that

$$\sum_{1}^{\infty} A_n \sin{(n\pi x)} \cos{(an\pi t)} \tag{10.12}$$

is a solution? If the coefficients were such that (10.12) makes sense and that we could differentiate twice under the summation sign, then (10.12) would provide a solution to the partial differential equation (10.1). Justification of such a step would require considerable analysis, beyond the scope of this course, but it will be easy to see what the coefficients *must be* if (10.12) is to be a solution of our problem.

First compute the Fourier sine series (the eigenfunction expansion) of the continuous function $f(x)$, the initial displacement. This is given by

$$f(x) = \sum_{1}^{\infty} c_n \sin{(n\pi x)}$$

where

$$c_n = 2\langle f(x), \sin{(n\pi x)}\rangle$$

$$= 4\left[\int_{0}^{1/2} x\sin{(n\pi x)}\,dx + \int_{1/2}^{1} (1-x)\sin{(n\pi x)}\,dx\right]$$

$$= \frac{8}{n^2\pi^2}\sin{\frac{n\pi}{2}}.$$

If $u(x, t)$ is defined by (10.12), then to satisfy the initial condition $u(x, 0) = f(x)$, it must be the case that

$$\sum_{1}^{\infty} A_n \sin{(n\pi x)} = \sum_{1}^{\infty} \frac{8}{n^2\pi^2}\sin{\frac{n\pi}{2}}\sin{(n\pi x)}$$

or that

$$u(x, t) = \sum_{1}^{\infty} \frac{8}{n^2\pi^2}\sin{\frac{n\pi}{2}}\sin{(n\pi x)}\cos{(an\pi t)}.$$

Note that not only is (10.1) satisfied, but the initial conditions (10.4) and (10.5) and the boundary conditions (10.1) and (10.2) are as well. This function, $u(x, t)$, represents the motion of the string as a function of position and time.

The reader should note the important role played by the eigenvalues. The parameter k first appeared rather mysteriously but then was fixed at one of a countable set of allowable values (eigenvalues) if there were to be a nontrivial solution—if the string were to vibrate. This eigenvalue multiplied by the number

a (which was a function of the type of string) fixed the frequency of the time-dependent part of the solution. There was a fundamental frequency, $2\pi a$, and multiples of it, $2\pi na$ (actually, only odd multiples for our particular problem). The solution was a combination of terms oscillating at this fundamental frequency and its multiples (harmonics). The length of the string—part of the boundary conditions—affected the frequency, while the nature of the initial displacement (an initial condition) affected the blending of the harmonics by affecting the coefficients of each term.

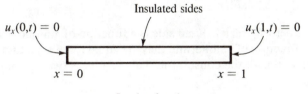

Figure 10.3   A one-dimensional rod with insulated ends.

The second application is the temperature distribution in a bar of finite length with insulated ends (see Figure 10.3). For convenience, we take the bar to be one-dimensional and of one unit length. Denote the temperature at time t and location x by $u(x, t)$. That the ends of the bar are insulated means that there is no heat flowing through them. This can be expressed mathematically as boundary conditions

$$u_x(0, t) = 0 \tag{10.13}$$

$$u_x(1, t) = 0, \qquad t > 0. \tag{10.14}$$

These conditions are sometimes called *zero flux conditions*. The initial temperature in the bar is also given as an initial condition, expressed by

$$u(x, 0) = f(x), \qquad 0 \leq x \leq 1. \tag{10.15}$$

As the temperature tries to equilibrate it does so according to a cooling law, which we model by the partial differential equation

$$u_t = ku_{xx}. \tag{10.16}$$

Equation (10.16) is called the *heat equation* or the *diffusion equation*. Its derivation can be found in many physics textbooks.

We seek to solve the initial boundary value problem [(10.13)–(10.16)]. As before, we attempt to do this by the separation of variables technique, that is, by attempting to find a solution of the form $u(x, t) = X(x)T(t)$. A basic existence and uniqueness theorem will tell us that if a solution of this form can be found, then it is the only solution. Substituting this function into (10.16) yields that

$$X(x)T'(t) = kX''(x)T(t)$$

or that

$$\frac{X''(x)}{X(x)} = \left(\frac{1}{k}\right)\frac{T'(t)}{T(t)}.$$

Proceeding as before, the left-hand side is a function of x alone, while the right-hand side is a function of t alone, and they are equal for every x and t; hence, they must be constant. Call the constant w. This produces two ordinary differential equations

$$X'' - wX = 0$$

$$T' - kwT = 0,$$

for which we need to derive the necessary boundary conditions.

To meet the condition that $u_x(0, t) = 0$ for $t > 0$ requires that $X'(0)T(t) = 0$ for $t > 0$, which will lead to the trivial solution if $T(t) = 0$ for all $t > 0$. Hence we impose the boundary condition $X'(0) = 0$. For similar reasons, $X'(1) = 0$ is appropriate. Thus we obtain a Sturm-Liouville problem,

$$X'' - wX = 0 \tag{10.17}$$

$$X'(0) = X'(1) = 0. \tag{10.18}$$

It is not difficult to argue that the parameter w must be real. (The assumption $w = u + iv$, $v \neq 0$ leads quickly to a contradiction.) We consider separately the possibilities that $w > 0$, $w = 0$, or $w < 0$. Suppose that $w > 0$. Then solutions of the differential equation are of the form $X(x) = c_1 \sinh(\sqrt{w}x) + c_2 \cosh(\sqrt{w}x)$. Imposing the first boundary condition leads immediately to the conclusion that $c_1 = 0$. Since $\sinh(x)$ has no zeros other than $x = 0$, it follows that $c_2 = 0$, and hence we have the trivial solution. Thus, since $w > 0$ is not possible, we make the change of parameter by letting $w = -\lambda^2$, and consider

$$X'' + \lambda^2 w = 0 \tag{10.19}$$

$$X'(0) = 0 \tag{10.20}$$

$$X'(1) = 0. \tag{10.21}$$

Solutions of (10.19) are of the form $c_1 \cos(\lambda x) + c_2 \sin(\lambda x)$. The first boundary condition leads to $c_2 = 0$, while fitting the second requires that $\lambda = n\pi$, or

$$X_n(x) = C_n \cos(n\pi x),$$

where C_n is arbitrary and $n = 1, 2, \ldots$. There is also a solution corresponding to $\lambda = 0$, namely $X_0 = C_0$, that satisfies the equation and the boundary conditions. With these values of λ we return to the T equation

$$T' + kn^2\pi^2 T = 0$$

to find that $T = B_n e^{-n^2\pi^2 kt}$. Thus a solution of the heat equation (10.16) and the boundary conditions (10.13) and (10.14) is given by (combining constants)

$$u_n(x, t) = A_n e^{-n^2\pi^2 kt} \cos(n\pi x), \qquad n = 1, 2, \ldots$$

and

$$u_0 = A_0.$$

Since the equation is linear, any linear combination of solutions of (10.16) is again a solution. A similar statement applies to the boundary conditions. Thus any linear combination of solutions of the boundary value problem will also be a solution. Hence we know that

$$u(x, t) = A_0 + \sum_1^N A_n e^{-n^2\pi^2 kt} \cos(n\pi x)$$

is a solution of (10.16), (10.13), (10.14). Without justification, we assume that

$$u(x, t) = A_0 + \sum_1^\infty A_n e^{-n^2\pi^2 kt} \cos(n\pi x) \tag{10.22}$$

is a solution as well. The remaining condition, (10.15), the initial condition, will be met by appropriately selecting the coefficients in the sum. As we expect, the theory of Fourier series will provide a mechanism for doing this.

To meet the initial temperature distribution in the bar it must be the case that

$$u(x, 0) = f(x).$$

Expand $f(x)$ in a Fourier cosine series:

$$f(x) = F_0 + \sum_1^\infty F_n \cos(n\pi x).$$

Equation (10.22) has the form, at $t = 0$,

$$u(x, 0) = A_0 + \sum_1^\infty A_n \cos(n\pi x).$$

Equating coefficients of this representation with that obtained in (10.22) for $t = 0$ yields the choice

$$A_0 = F_0 = \langle f, 1 \rangle$$
$$A_n = F_n = 2\langle f, \cos(n\pi x) \rangle.$$

With this choice of coefficients, (10.22) now satisfies the heat equation, the boundary conditions, and the initial condition. The presence of the negative exponential ensures that each term in the sum tends to zero and, in fact, does so sufficiently rapidly that as t becomes large, the temperature tends to a constant (to A_0, the mean value of f). This meets with our physical intuition that the heat, and thus the temperature, would distribute uniformly in the bar.

EXERCISES

1. Show that the boundary value problem

   $$y'' - ky = 0, \qquad k > 0$$
   $$y(0) = y(1) = 0$$

   has only the trivial solution. (This justifies the choice of $-k$ for the parameter in the wave equation when the variables are separated.)

2. Show that $f(x + at)$ and $f(x - at)$ are solutions of (10.1) for any twice-differentiable function f.

3. Make the change of variables $z = x + at$ and $w = x - at$, and show that if $u(x, t)$ is a solution of (10.1), then $u(z, w)$ is a solution of $u_{zw} = 0$. (*Hint:* Use the chain rule.)

4. Construct a solution of $u_{zw} = 0$. (*Hint:* The argument begins that $(u_z)_w = 0$ implies that $u_z = g(z)$, where $g(z)$ is an arbitrary differentiable function of z. Continue, to obtain $u(z, w) = G(z) + H(w)$.)

5. Suppose that $u(x, t)$ is a solution of (10.1); $u(x, 0) = f(x)$, where f has two derivatives; and $u_t(x, 0) = 0$. Show that $u(x, t) = (f(x + at) + f(x - at))/2$. (*Hint:* Use Exercises 3 and 4.)

6. Find the solution to (10.1)–(10.5) when $f(x) =$

   (a) x

   (b) $\sin(\pi x)$

   (c) $x(1 - x)$

7. Suppose that a rod with insulated ends is one unit long and is made of material such that, in (10.15), $k = 1$. If the initial temperature distribution is $f(x) = x$, find the temperature as a function of x and t.

8. Suppose, in Exercise 7, that the initial temperature distribution is $f(x) = x$ for $0 < x < 1/2$ and $f(x) = 1 - x$ for $1/2 < x < 1$. Find the temperature as a function of x and t. Find $\lim_{t \to \infty} u(x, t)$. (This is called the *steady-state temperatue* if the limit exists.)

9. If the ends of the bar are held at fixed temperatures A and B instead of being insulated, the resulting model is of the form

$$u_t(x, t) = ku_{xx}(x, t)$$

$$u(0, t) = A, \qquad u(1, t) = B, \qquad t > 0$$

$$u(x, 0) = f(x), \qquad 0 < x < 1.$$

Assume that separation of variables will work and find the form of the solution.

10. Suppose that one end of the bar is held at a constant temperature and that the other end is insulated. What is the form of the boundary conditions?

11. Assume that separation of variables works for Exercise 10. Derive the form of the solution, given an additional initial condition.

11. Nonlinear Boundary Value Problems

All of the boundary value problems considered in this chapter have involved linear differential equations and linear boundary conditions. Although the study of nonlinear boundary value problems is beyond the scope of this course, it would create an erroneous impression to leave the subject without at least an indication of the difficulties that arise when the problems become nonlinear. The subject of nonlinear boundary value problems and, in particular, nonlinear Sturm-Liouville problems is a research topic of recent interest in mathematics, one that developed to explain very practical phenomena. We first discuss a simple example to show how things go wrong when the differential equation is nonlinear and then illustrate the kind of result that is possible for nonlinear Sturm-Liouville problems and compare this with the corresponding linear result developed earlier.

Consider the boundary value problem

$$y'' + |y| = 0 \tag{11.1}$$

$$y(0) = 0 \tag{11.2}$$

$$y(b) = B, \tag{11.3}$$

where b and B are to be treated as parameters. The behavior of the solutions will vary as these quantities vary. Recall the formal definiton of absolute value,

$$|y| = \begin{cases} y, & y \geq 0 \\ -y, & y < 0. \end{cases}$$

A solution of (11.1), (11.2) and $y'(0) = m$, $m > 0$ is also a solution of the differential equation

$$y'' + y = 0, \tag{11.4}$$

as long as $y(t)$ remains positive. A solution of (11.1), (11.2), $y'(0) = m$, $m < 0$, satisfies

$$y'' - y = 0, \tag{11.5}$$

as long as $y(t)$ remains negative. Of course, if $m = 0$, then $y(t) = 0$ is a solution for all t. All solutions of (11.4) and (11.2) are of the form $C \sin(t)$ for some constant C, while all solutions of (11.2), (11.5) are of the form $C \sinh(t)$ for some constant C. We will make use of these facts to construct solutions of the nonlinear differential equation by piecing together solutions of the linear equations.

We can solve the boundary value problem (11.5), (11.2), (11,3) for $b > 0$ and $B < 0$ by making a judicious choice of the constant C, namely by taking the function

$$y(t) = B[\sinh(b)]^{-1} \sinh(t). \tag{11.6}$$

(Check it.) Further, this is the only solution for this linear boundary value problem. Since $B < 0$, $y'(0) < 0$, and the above function is negative for all $t > 0$ (sinh (t) has no positive zeros). Thus, by the definition of absolute value, the function $y(t)$ given by (11.6) is also a solution of the nonlinear boundary value problem (11.1)–(11.3) for $B < 0$ and any $b > 0$.

If $b < \pi$, and $B > 0$, the boundary value problem (11.4), (11.2), (11.3) has the unique solution

$$y(t) = B[\sin{(b)}]^{-1} \sin{(t)}. \tag{11.7}$$

With b and B restricted as before, this function is a solution of the nonlinear boundary value problem (11.1)–(11.3) as well. If $B = 0$, then $y(t) = 0$ is the unique solution of (11.4) and the boundary conditions (11.2) and (11.3) if $b < \pi$, while $y(t) = 0$ is also the solution of (11.5) and the boundary conditions and it is unique for all $b > 0$.

Now suppose that $B \geq 0$ and $b \geq \pi$. Every solution of (11.1) and (11.2) with $y'(0) > 0$ has a zero at π (since every solution of (11.4) and (11.2) is a multiple of $\sin{(t)}$) and has $y'(\pi) < 0$. Once this function crosses zero it is no longer a solution of (11.1). Moreover, as previously noted, any solution of (11.1) that is zero at a point and has a negative derivative at that point is a solution of (11.5) for as long as it is negative. However, a solution of (11.5) that is zero at a point t_0 and has a negative derivative is negative for all future values of t. (Indeed, it is of the form $C \sinh{(t - t_0)}$ for $C < 0$.) We reach two immediate conclusions. If $b = \pi$ and $B = 0$, then there are infinitely many solutions of (11.1)–(11.3), one for each choice of initial slope. If $b > \pi$ and $B > 0$, there is no solution of (11.1)–(11.3). In more formal terms, existence fails; the reader might want to reread the existence theorem for boundary value problems discussed at the end of Chapter 3 to better understand the significance of the hypotheses of that theorem.

If $b > \pi$ and $B < 0$, the situation is much more interesting. As noted before, one solution is given by (11.6). A second solution may be constructed that starts at $t = 0$ with a positive initial slope, has a zero at $t = \pi$, and then reaches B at $t = b$. We construct the solution in two pieces,

$$y(t) = \begin{cases} c_1 \sin{(t)}, & 0 \leq t \leq \pi, \\ c_2 \sinh{(t - \pi)}, & \pi \leq t \leq b. \end{cases} \tag{11.8}$$

To satisfy (11.3), we must choose $c_2 = B[\sinh{(b - \pi)}]^{-1}$. There remains the task of matching the derivatives of the two parts at the zero of the function. This can be accomplished by choosing $c_1 = -c_2$. With this choice of constants, the left- and right-hand limits of the function $y'(t)$ at $t = \pi$ are c_2. Further, the left- and right-hand limits of $y''(t)$ at $t = \pi$ are zero. As a consequence, the function defined by (11.8) is a solution of (11.1) for $0 < t < b$. Thus there are two solutions to the original boundary value problem, (11.6) and (11.8). With a little additional effort we can show that there are exactly two solutions in this case.

This discussion may be summarized as follows.

1. If $b < \pi$, the problem (11.1)–(11.3) has a unique solution.

2. If $B > 0$ and $b > \pi$, the problem has no solution.

3. If $B < 0$ and $b > \pi$, the problem has two solutions.

4. If $B = 0$, there is a unique solution of the problem for every value of b except $b = \pi$, where there are infinitely many solutions.

A schematic is shown in Figure 11.1. Note how different this example is from the linear problems discussed early in this chapter. This example shows that without further conditions, we can anticipate the existence of a solution only if the distance between the points at which the boundary conditions are placed is small. In the problem above this corresponds to $|b|$ being small.

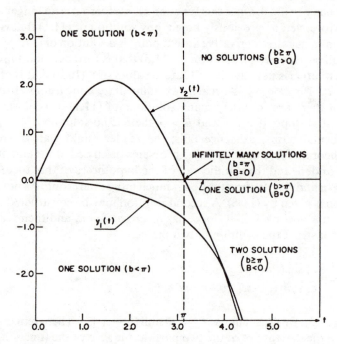

Figure 11.1 Schematic of the solutions of (11.1)–(11.3). (Reproduced from P. B. Bailey, L. F. Shampine, and P. Waltman, *Nonlinear Two Point Boundary Value Problems.* (New York: Academic Press, 1968), 9.)

We turn now to the nonlinear Sturm-Liouville problem. We illustrate the problem by considering a relatively simple equation and boundary conditions

$$y'' + [\lambda - Q(t)]y = yf(t, y) \tag{11.9}$$

$$y(0) = 0 \tag{11.10}$$

$$y(1) = 0. \tag{11.11}$$

The results described are fairly typical of the spirit of the results for nonlinear problems. If $Q(t)$ is continuous and $f(t, y) = 0$ for all t and y, then a linear eigenvalue problem,

$$y'' + [\lambda - Q(t)]y = 0 \tag{11.12}$$

$$y(0) = 0 \tag{11.13}$$

$$y(1) = 0, \tag{11.14}$$

results, which is a special case of those considered previously. From Theorem 5.1 it follows that there is a sequence of eigenvalues

$$\lambda_0 < \lambda_1 < \lambda_2 < \cdots$$

and from Theorem 6.2 it follows that an eigenfunction corresponding to λ_n has exactly n zeros.

We will assume that f is a continuously differentiable function of all of its arguments, that $f(t, y) > 0$ for $0 < t < 1$, $y \neq 0$, and that $f(t, 0) = 0$. In particular, then, $f(t, y)$ is small for small values of y. The form of equation (11.9) and the boundary conditions may be generalized and the conditions above on f may be considerably weakened. Since we are only illustrating the theory, we will keep this simple form. A typical local theorem is as follows.

THEOREM 11.1

Let the above hypotheses on Q and f hold. Then for each $n = 0, 1, 2, \ldots$ there is an interval $I_n = (\lambda_n, \lambda_n^*)$ such that for $\lambda \in I_n$, (11.9)–(11.11) has at least two solutions with exactly n zeros in $(0, 1)$.

The theorem is best understood if it is contrasted with the linear result. Recall that for the linear Sturm-Liouville problem the eigenvalues formed a discrete set, and that any multiple of an eigenfunction was again an eigenfunction. This last property clearly cannot hold without further restrictions on f. Moreover, the special values of the parameter λ now form an interval whose left-hand endpoint is the eigenvalue of the corresponding linear system.

The discussion above says nothing about the size of the interval or the magnitude of the solutions. Understanding what happens to these two quantities requires considerable work. However, with further limitations on growth of the functions f, we can change the conclusion of the theorem to make $I_n = (\lambda_n, \infty)$.

REFERENCES

Advanced texts covering boundary value problems include

> Coddington, E. A., and N. Levinson. *Theory of Ordinary Differential Equations.* New York: McGraw-Hill, 1955.

> Atkinson, F. V. *Discrete and Continuous Boundary Value Problems.* New York: Academic Press, 1964.

A very readable intermediate-level presentation can be found in the following. (Some portions of Chapter 4 follow this approach.)

> Birkhoff, G., and G.-C. Rota. *Ordinary Differential Equations.* New York: John Wiley & Sons, 1978.

The construction of Green's functions, and many other topics in boundary value problems, can be found at the intermediate level in

> Stakgold, I. *Boundary Value Problems of Mathematical Physics.* New York: Macmillan, 1967.

The following is the classical reference on eigenfunction expansion. This approach requires an understanding of complex variables.

> Titchmarsh, E. C. *Eigenfunction Expansions Associated with Second Order Differential Equations.* Oxford: Oxford University Press, 1962.

The separation of variables technique can be found in any of the "advanced mathematics for engineers" textbooks. See also

> Churchill, R. V. *Fourier Series and Boundary Value Problems.* New York: McGraw-Hill, 1963.

The example of a nonlinear boundary value problem appears in

> Bailey, P. B., L. F. Shampine, and P. Waltman. *Nonlinear Two Point Boundary Value Problems.* New York: Academic Press, 1968.

Theorem 11.1 is a special case of one result in

> Crandall, M. G., and P. Rabinowitz. "Nonlinear Sturm-Liouville Eigenvalue Problems and Topological Degree." *J. Math. Mech.* 19 (1970): 1083–1102.

See also

> Macki, J. W., and P. Waltman. "A Nonlinear Sturm-Liouville Problem." *Indiana Univ. Math. J.* 22 (1972): 217–25.

The topology reference for completion of a metric space is

> Dugundji, J. *Topology.* Boston: Allyn and Bacon, 1970.

Index